SAY IT WITH FIGURES

ABOUT THE AUTHOR

Hans Zeisel is Professor of Law and Sociology Emeritus at the University of Chicago where he pioneered the application of social science to law. Earlier he had a distinguished career in public opinion and market research; in 1980 he was inducted into the Market Research Hall of Fame. He has written on a wide variety of topics, ranging from research methodology and history to law enforcement, juries, and Shakespeare. He has been elected fellow of the American Statistical Association and of the American Association of the Advancement of Science.

SAY IT WITH
FIGURES

SIXTH EDITION

HANS ZEISEL

1817

HARPER & ROW, PUBLISHERS, New York
Cambridge, Philadelphia, San Francisco,
London, Mexico City, São Paulo, Singapore, Sydney

To the Memory
of
Paul Felix Lazarsfeld

Designer: C. Linda Dingler

Library of Congress Cataloging in Publication Data

Zeisel, Hans.
  Say it with figures.
  Includes index.
  1. Social sciences—Statistical methods. 2. Statistics. I. Title.
HA29.Z4   1985   001.4'22   84-48207
ISBN 0-06-181982-4   85 86 87 88 89 10 9 8 7 6 5 4 3 2 1
ISBN 0-06-131994-5 (pbk.)   85 86 87 88 89 10 9 8 7 6 5 4 3 2 1

Contents

Contents

Preface

The revisions in this sixth edition of *Say It with Figures* are of three kinds. Although the timeliness of the examples has never been a primary aim, I have freshened up some of them. Second, many topics dealt with in the book have developed new niches and perspectives, which now have their place in the text. Third, but not least, Part II has been substantially reshaped by giving greater prominence to the controlled randomized experiment, because of its growing practical importance and its essential role as a guide to the expanding field of quasi-experimental techniques. The change was made in the spirit of Donald Campbell's vision of the experimental society in which the experiment will be a steady, critical, and thereby encouraging companion to reform.

I have added a new chapter on regression analysis. The advent of the computer has made this technique a widely if not always wisely used tool. I had begun to wonder whether a presentation of regression, its uses and abuses, might not serve a useful purpose. All doubts were removed when my revered friend Judge Marvin Frankel, learning that I was revising the book said to me, "Be sure that after I have read it I will know what regression analysis is." Although I have gone my own way in this effort, I have with the permission of the authors borrowed several graphs from a splendid statistical text, simply because I could not possibly improve on them.[1]

There is an anomaly about *Say It with Figures* to which I should like to draw attention. After many years of striving to present statistical data well, I have concluded that only statisticians or statistically curious people read statistical tables. For the ordinary reader, tables are an unwelcome chore, encouraging the inclination

---

[1] David Freedman, Robert Pisani, and Roger Proves, *Statistics*, (New York: Norton, 1978).

to pass them over. In my latest book I have drawn the consequences from that insight and presented most of the data in graph form. I wrote there, "tables make difficult reading. In their place are graphs which convey the statistical evidence without pain and perhaps even with pleasure."[2] I regret that I cannot treat the readers of *Say It with Figures* with equal consideration. But graphs are based on tables, and the care and feeding of such tables remains one major concern of this book. There are nevertheless some thirty odd graphs in this text which my friend Darryl Beck has put into elegant shape.

Over the years, in many countries, I have met people who remember reading *Say It with Figures* with pleasure, mainly for two reasons: it discusses problems ever present in the statistician's workday that are not discussed elsewhere; and it clarifies some of the basic problems common to all efforts, however advanced, to distill cause-and-effect relations. This revised edition retains that twofold function. The technical literature on the study of causes and effects has grown considerably, and in a different book that progress would have to be taken note of; I have tried to avoid the overlap with such literature. This self-limitation also explains why I have cited only a few authors, even though I am indebted to many.

In the forewords to earlier editions I thanked friends for their direct and indirect contributions to the gestation of this book: Cuthbert Daniel, Edward Suchman, C. Wright Mills, Ruby Taylor, Philip Ennis, Herta Herzog, and Ilse Zeisel, all of whom were then part of the Bureau of Applied Social Research at Columbia University; Matilda White and Raymond Franzen of the old Market Research Company of America; Marion and Virginia Harper, and Esther Mulder of McCann-Erickson; Lotte Radermacher of the Vienna Institute for Psychological Market Research; and my late friend and colleague at the University of Chicago,

---

[2] *The Limits of Law Enforcement*, (Chicago: The University of Chicago Press, 1983).

Harry Kalven, Jr., with whom I shared many happy years of saying it with figures to our colleagues in the law.

Cuthbert Daniel, Herta Herzog, and Ilse Zeisel have once again helped with this sixth edition. I also want to thank William Kruskal for his continuous transmission of nourishing critical comment.

Were life to go on forever, Paul Lazarsfeld would have written the introduction to this sixth edition. As it is, it must suffice to reprint his introductions to the first and the last edition. They tell of the origin of *Say It with Figures* and delineate the place we hoped it would occupy in the library of research methodology. Many of those hopes were fulfilled. The present revision is meant to allow the book to retain its happy place.

I dedicated the first edition of this book to the memory of Klementine Enslein, my incomparable and beloved grammar school teacher, who a long time ago in Vienna taught me to enjoy figures. *Say It with Figures* I trust will continue to make its contribution to such enjoyment, the main root of all intellectual endeavor.

University of Chicago   H.Z.
1984

Introduction
to the Fifth Edition

It is worthwhile to reflect on the success of a research text that enters its fifth American edition and has been translated into a number of foreign languages. Having watched Professor Zeisel's plan from the beginning, it is not difficult for me to explain its success and to document the explanation.

The present book deals with ways to study human affairs in a variety of fields—law, consumer choices, economics, public opinion. But it is organized around basic methodological ideas. Twenty-five years ago, they had been tentatively laid out, today they have become classics. Look at the new *Encyclopedia of the Social Sciences*, at the topics that cannot be found in the earlier edition. You will find there extensive entries on reason analysis, panels, and cross-tabulation. The authors practically parallel the outline of *Say It with Figures*, and, incidentally, many of them are alumni of Columbia's Bureau of Applied Social Research, of which Dr. Zeisel has been a valuable consultant for these many years. So this book has helped in developing a structure that today dominates research and practice.

But no formal structure can survive long if it does not come alive in concrete examples, judiciously chosen to illustrate and clarify specific points. And here is a second reason for the longevity of *Say It with Figures*. From the literature which the author has continually watched and from his own ever broadening research experience, he has been replacing old examples with new ones, and has widened their range. Especially noteworthy in this new edition is the inclusion of studies from the field of law, in which Zeisel himself has made groundbreaking investigations, and, in the chapter on indices, the new bridges to the work of the economists.

It is fine to maintain a tradition, but it is also dangerous; new ideas might be overlooked. And here is the third merit that accounts for the book's success. In every edition, attention is paid to ways in which the profession, or the author himself, has added variations to the basic themes, or opened new vistas. First there were chapters on cross-tabulation; later the chapters on reason analysis were added. Now, in a new chapter, the integration of the two approaches broadens the field of empirical causal analysis. This new chapter, on the triangulation of proof, illustrates the possibilities of increasing the power of our analysis by looking for confluence of evidence from independent sources.

Thus this text gives meaning to the often abused notion of interdisciplinary work. In general, problems don't get solved by asking, say, a business research man and a sociologist to work together. What links the conventional disciplines are procedures of inquiry, integrated in the professional training of that younger generation that is not any longer concerned with terminological distinctions. The government, the public-health expert, the politician, the sales manager, the leftist organizer—they all wait for a new profession: the research expert on human affairs, who combines a variety of skills, just as the medical doctor combines training in basic sciences with clinical imagination. Conventional higher education is only slowly recognizing this need; often it puts blocks into the road of progress. If *Say It with Figures* keeps on saying it, it will help to free the passage. The next generation of students may yet see the road open to an ever more expert training in the use of social inquiry, for an ever broadening range of decision problems.

Columbia University   Paul F. Lazarsfeld
1968

From the Introduction
to the First Edition

Modern social life has become much too complicated to be perceived by direct observation. Whether it is dangerous to take an airplane, whether one kind of bread is more nourishing than another, what the employment chances are for our children, whether a country is likely to win a war—such issues can only be understood by those who can read statistical tables or get someone to interpret them.

The very complexity of social events requires a language of quantity. And yet, one who has observed students of the social sciences knows how many have trouble when they want to "say it with figures." I do not believe that this is due to any inherent difficulties. It stems rather from a certain inconsistency in our statistical training. A personal reminiscence of how the present book developed might help to clarify this thought.

After World War I, Professors Karl and Charlotte Bühler directed the Department of Psychology at the University of Vienna. Under their leadership, it became a center for the application of psychology to social problems. We were continuously confronted with topics like these: how do young people acquire "work consciousness" and finally vocational maturity? How does the behavior of parents affect the relationships among siblings? By what criteria do old people, looking back over their life, decide whether it has been meaningful? Is the morale of unemployed men better preserved by dole or by work relief? The questions were the outgrowth of systematic theories about the course of human life and its relation to the social system. But the answers were sought through concrete material: a large collection of diaries kept by young people, carefully recorded observations of family situations, detailed interviews with residents of old age homes, surveys in unemployed communities, and so on.

As an assistant to the Bühlers, I was in charge of training students to handle such material. Little precedent for this task could be found in the tradition of the social sciences. The categories were more complex than those usually treated by quantitative methods; because they were what is called today "qualitative attributes," no standard correlation techniques could be used. Furthermore, the goal was not to find isolated relationships. The results had to hang together, each as part of a consistent whole. This situation led, not to the development of new formulae, but to a kind of empirical work in which qualitative analysis is guided by conceptual schemes and in which each empirical procedure is scrutinized as to its logical implications. After coming to this country, I realized how much it would have helped had we known more about the statistical methods developed by American scholars. In my teaching here, however, I have found that the Viennese tradition is also worth preserving.

Columbia University's Department of Sociology has a special division for the research training of its students, the Bureau of Applied Social Research. Since its inception, we have been bothered by a gap in the available literature.

There is a no-man's land between everyday language and systematic statistical procedures. The empirical research man may get and may give training in formal statistics in our colleges. But those who have observed research projects in government and industry have noticed that these techniques sometimes fail to solve his problems. What is needed equally is an intelligent grasp of what figures stand for and what they can be used to express. The same young research worker who has learned very well to compute a probable error falls down when he is expected to interpret a large set of simple percent figures, and he fails still further when he must explain his interpretation to a layman. It is usually assumed that no rules exist in this twilight area; the intuition of the practitioner is supposed to provide a way.

This, I feel, is a misconception and an impediment to the progress of social research. In an effort to overcome it, it seemed desirable to present explicitly, and with detailed discussion of examples, some of the procedures which are treated rather casually elsewhere. For, wherever it is possible to "codify," there is a better opportunity to teach as well as to learn. The present book is a first step in the direction of codification. Dr. Hans Zeisel and I have worked together here and abroad, and he is well acquainted with the trend of thought expressed in the preceding paragraphs. The Bureau of Applied Social Research has turned over its files of studies and training materials to him and to this he has added his own experience as research officer of a large commercial agency. Dr. Zeisel has written a text which should stimulate the student, as well as the practical research man, to see the logic behind familiar research procedures and to develop new techniques from such improved understanding.

The examples throughout the book were chosen from a wide variety of fields. A special effort has been made to intermingle materials from market research, sociology and psychology. There is no *logical* difference between the study of voting or of buying. In each of these areas, the final goal is the discovering of regularity in social life. In most cases the examples have been taken from actual research studies; where the didactic purpose seemed to require it, the data were simplified so as to bring out their logical implications.

The reading of the book does not require any preliminary knowledge of quantitative methods. In many places a more systematic approach was sacrificed for the sake of simplicity. For this, and numerous other reasons, the text is by no means a final system. It is an effort to stimulate a certain way of looking at research material, of analyzing and presenting it. This publication is only a beginning in which we hope many others will join, contributing their observations, problems, and results.

1947   Paul F. Lazarsfeld

PART I

THE PRESENTATION
OF NUMBERS

The relevance and precision of the numbers that make up our statistical tables are forever increasing because our analytic tools are getting better. Yet at the point where we present these numbers, we are often careless. Statistical tables do not always reflect the lucidity of our analysis and therefore fail in their prime purpose of communicating their message.

The first six chapters of this book try to improve this situation by drawing attention to difficulties and solutions in the feeding and care of statistical tables.

Chapter 1 deals with the function and power of percent figures. This ubiquitous and powerful device of presenting numerical results, it turns out, must be handled with circumspection lest the message they are to convey be distorted or lost.

Chapter 2 deals with the visual difficulties of presenting statistical tables. The chapter begins with problems posed by percent figures and ends with a brief excursion into graphic solutions. We are all familiar with the translation of simple numerical relations into bar charts and similar clarifying devices. There is also a growing literature on the graphic presentation of complex relations by ever more sophisticated graphic devices.[1] The excursion at the end of Chapter 2 draws attention to the possibility of presenting complex concepts in relatively simple graphs, a task that cannot be solved routinely.

[1] For instance, Edward R. Tufte, *The Visual Display of Quantitative Information* (Cheshire, CO: Graphics Press, 1983).

Chapter 3 covers a problem which, if not properly resolved, has marred many statistical tables: the direction in which percent figures should be run. The problem has achieved new actuality through the computer printouts that usually have them run in both directions.

Chapter 4 deals with the minor but at times major difficulties caused by the Don't Knows and No Answers that often form the bottom line of statistical tables.

Chapter 5 offers solutions for the vexing problem of how to present in readable form relations between more than two factors. The trick is to represent one of the dimensions by a single number.

Chapter 6, on Indices, expands on such an initial step by showing how even more complex concepts can be represented by a single number and thereby become potentially subject to further statistical analysis. The formula that transforms the concept into a single figure both reflects and in turn helps to define the concept for which it stands.

1

The Function of Percent Figures

The primary purpose of percent figures is to clarify the relative size of two or more numbers. They achieve this goal in two ways. They designate one number, usually their sum, as the base, and translate it into the figure 100; the other numbers are then translated in proportion into numbers smaller than 100 which allows their various relations to be perceived more easily. Something, of course, is lost, albeit not entirely. The numbers themselves have disappeared; but they can be reconstituted by multiplying the base number by the respective percentages.

FACILITATING COMPARISONS

Percent figures are particularly useful when ratios are to be compared that have different base numbers. Consider the world production of automobiles by country for the years 1960 and 1980 (Table 1-1).

In the 20-year interval, the United States' production remained virtually the same. Great Britain's production declined markedly. All other countries increased their output; one country—Japan—increased it spectacularly from 482,000 cars to 11,043,000. World production more than doubled; for that reason it is not easy to see from this Table 1-1 the changes in the *relative* standing of these competing countries. Table 1-2 shows the changes with greater clarity.

By merely holding its output still, the United States' share of the world market dropped from 48.3 to 20.9 percent. Great Britain's share, because its production dropped in an expanding market, fell from a respectable 11.1 percent to a mere 3.4. The sensational gain is Japan's, which increased its share from 2.9 to 28.8 percent and removed the United States from its position of leadership.

TABLE 1-1

Changes in World Production of Automobiles

(In 000)

	1960	1980
United States	7,905	8,010
West Germany	2,055	3,878
Great Britain	1,811	1,313
Other Europe	2,200	6,470
Eastern Bloc	1,017	4,254
Canada, Brazil, Mexico, Australia	907	3,393
Japan	482	11,043
Total	16,377,000	38,361,000

Source: R. L. Polk & Co.

TABLE 1-2

World Production of Automobiles (Market Shares)

	1960 %	1980 %
United States	48.3	20.9
West Germany	12.5	10.1
Great Britain	11.1	3.4
Other Europe	13.5	16.9
Eastern Bloc	6.2	11.1
Canada, Brazil, Mexico, Australia	5.5	8.8
Japan	2.9	28.8
Total percent	100.0	100.0
Number	(16,377,000)	(38,361,000)

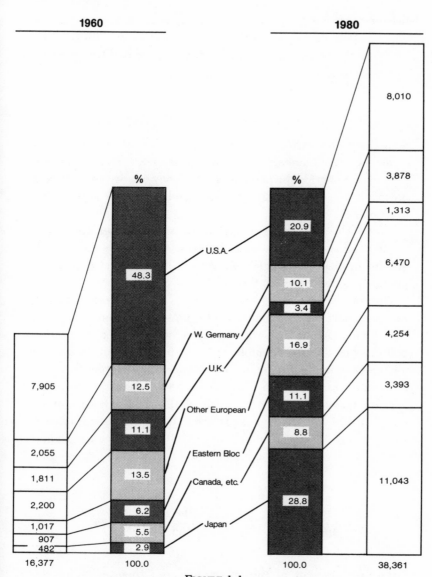

FIGURE 1-1
World Production of Automobiles, 1960 v. 1980

Figure 1-1, by combining Tables 1-1 and 1-2, depicts the principle of the transformation of numbers into percents. By equalizing the bases of both number columns to 100, and by reducing the other numbers proportionally, we find that comparison is facilitated.

DEEMPHASIZING THE NUMBERS

Something, of course, is lost in the course of the percent transformation: the numbers themselves disappear, unless they are preserved alongside the percent figures. They can be reconstituted, however, by multiplying the base number by the appropriate percentage.

Most of the time, these numbers from which the percent figures are derived are important in themselves. Automobile manufacturers, for instance, although interested in their market shares, are primarily concerned with the number of cars produced.

There are situations, however, in which only the percent figures have meaning and the underlying numbers have none. Such is the case in most survey work. Consider, for instance, a preelection poll in which 796 of a sample of 1,500 prospective voters voted for the Democratic candidate and 704 for the Republican. Taken by themselves, these two numbers have no meaning whatsoever, because they are partly the result of the arbitrary decision to sample 1,500 voters, a choice prompted by considerations of costs and sampling error. The numbers 796 and 704 become meaningful only in their relation to one another, more precisely, in relation to their sum.

$$\frac{796}{1,500} = .53 \text{ or } 53\% \qquad \frac{704}{1,500} = .47 \text{ or } 47\%$$

Fifty-three percent of the voters declared themselves for the Democratic candidate and forty-seven percent for the Republican.

Mathematically, the expressions 796/1500 and 53 percent are equivalent, but 53 percent is the simpler and therefore the preferable expression.

CAUSAL IMPLICATIONS

The percent transformation will occasionally do more than make relations more transparent; percent figures can convey causal connections that may or may not be warranted. Suppose that company A increased its sales volume from one year to the next from $10 million to $20 million, and suppose that company B, a larger competitor of A, increased its sales during that same year from $40 million to $70 million, and we are asked to compare the sales progress of the two companies. Two such comparisons can be made:

Comparison I

A increased its sales by $10 million. B increased its sales by $30 million, three times the increase of A.

Comparison II

A increased its sales by 100 percent. B increased its sales by 75 percent, three-fourth the increase of A.

Strictly read, the two comparisons do not contradict one another, but comparison I gives the impression that company B did better than did company A; comparison II gives the opposite impression.

The point turns on the ambiguity of the term "better." Suppose that we choose a yardstick that allows the better-managed company to come out ahead. We must then consider whether the greater dollar increase of company B was primarily due to its better management or merely to its having been the larger company to start with. The choice, of course, will not always reflect merit;

those who want to color the picture in favor of company B may prefer the first reading; those who want to show how well A did will prefer the second one. But which is the correct version?

By expressing the increase as a percentage of the company's sales volume, we imply that it would not be fair to compare the dollar sales increase of a big company with that of a small one; the proper comparison would focus on the relative—the percent increase. The handicap resulting from company A's smaller size is neutralized by our expressing each company's increase as a percentage of the company's sales volume at the beginning of the period. The underlying assumption is that had the two companies operated during that year under equally favorable circumstances and under equally good management, both would have increased their sales by the same percentage. If the percent increase of one was greater than that of the other, this signifies that its management was better or at least luckier.

The dollar comparison, on the other hand, would be appropriate if there were little or no initial advantage in starting out as a larger company; if the two companies, for instance, were construction firms that obtained business from a small group of customers who base their choice on the reliability and quality of performance, without regard to size. If we assume that starting out as a larger company does *not* bring an initial advantage, we will accept the raw dollar figures as the appropriate performance measure and will reject the percent comparison.

The problem itself is an old one and has been discussed by no less a scholar than Galileo. It came up in this form: a horse, the true value of which was 100, had been appraised by two experts, one estimating its value at 10, the other at 1,000. The question was, who deviated more from the true value? Galileo thought both equidistant, since 1,000:100 = 100:10. Other participants in the learned dispute considered 1,000 more out of line than 10, the difference being 900 in the one direction and 90 in the other.[1]

---

[1] "Letters Concerning the Value of a Horse," *Opere di Galileo* (Florence), vol. XIV, 1855, pp. 231, 284.

TABLE 1-3
*Percent of Homes Burglarized
in County X*

1978	1979	1980	1981	1982
2.3	3.2	4.4	5.3	6.4

PERCENTS AND PERCENTAGE POINTS

If the change from one point of time to another is under scrutiny, it can be measured either in absolute numbers or in percents, whereby the first point in time is designated as the 100 percent base. This simple option is obscured if the percent measure has become the standard measure and we are confronted with a time series of percentages as in Table 1-3.

It would be correct to say that the number of homes broken into increased each year by approximately one percentage point. But if we want to know by how much the likelihood of being burglarized increased from one year to the next, we must perform a different computation.

WRONG END UP

Such translation from percentage point to percentages is particularly important if the percentage point figures which form the base data are small, as in Table 1-4. That translation acquired importance in a debate that ensued some time ago over how the insurance industry should treat automobile drivers who were poor safety risks and not able to obtain insurance.

In most states such drivers are assigned to a special pool in which all insurers have to share. At one time, however, an insurance analyst argued against the need for such a pool procedure by pointing at the numbers given in Table 1-5.

He asked, with respect to the tiny difference between 96.6 and 95.9, "Is the game worth the candle?" To which another analyst

TABLE 1-4

Percent Increase of Home Burglaries
over the Previous Year

1979 over 78	1980 over 79	1981 over 80	1982 over 81
$\dfrac{3.2}{2.3}$	$\dfrac{4.4}{3.2}$	$\dfrac{5.3}{4.3}$	$\dfrac{6.4}{5.3}$
+39%	+38%	+20%	+21%

correctly replied, "What matters is not the proportion of risk-free cases—they do not cost anything—but the cases in which claims were made. If the assigned cases are included, the loss ratio (the crude measure for the payout) increases from 3.4 to 4.1 percentage points, that is, by 21 percent, a difference that may be worth the candle."

TABLE 1-5

Comparing the Insurance Experience for Accepted Drivers
with That for All Drivers, including Those
Normally Assigned to the Reject Pool

Bodily Injury	Risks Accepted Selectively %	If All Drivers Were Accepted %
No damages claimed	96.6	95.9
Damages claimed	3.4	4.1
Total	100.0	100.0

Not only were the risk-free percentages the wrong figures to look at; in addition, the percentage point difference required translation into the appropriate percent difference.

THE CEILING EFFECT

So far we have discussed the possibility that the larger initial size may provide an advantage for future growth. There is also the obverse possibility that the larger initial size may be a handicap

TABLE 1-6

Readership of Three Advertisements Printed
in Black and White and in Color

(Percent of Readers of That Publication)

Advertisement	Black and White	Color
A	42	52
B	23	37
C	16	32

for growth because a countervailing force, the ceiling effect, makes growth increasingly difficult.

Suppose that some individuals go fishing with a net in a pond that is fairly full of fish. Each member of the group, in sequence, will try his or her skill by casting the net; the size of the catch will be the measure of each person's skill. This would be an unfair measure, because the earlier contestants will have an unmerited advantage. The first sweep with the net will yield a certain number of fish; that first catch, however, will have reduced the density of the fish in the pond and thereby put the next fisherman at a disadvantage. In this situation, neither the count of the fish nor their percent transformation will offer a fair comparison.

The paradigm applies to many constellations. Consider efforts to measure the effectiveness of advertisements. Readership of an advertisement is usually measured by the number of people who have read it, expressed as a percentage of all people who opened a particular magazine or newspaper. This yardstick has been used to measure the power of individual advertisements, as well as to compare different types of advertisements, in order to learn, for instance, how readership of an advertisement is affected by shifting from black and white to color.

Table 1-6 presents the readership data for three advertisements that had appeared both in black and white and in color. The question arose about whether these data allowed any general conclusion concerning the increase in readership to be expected

TABLE 1-7

Percent Increase in Readership from Black and White to Color

Adver-tisement	From	To	Measured in Percentage Points	Measured as Percent Difference Black and White = 100%
A	42	52	+10	+24
B	23	37	+14	+61
C	16	32	+16	+100

if an advertisement were to be produced in color instead of in black and white.

There are the two obvious ways of comparing the increase of readership in color over black and white—in absolute numbers, which in this case represent percentage points, or as percent increase of the color readership over that of the black and white.

Neither method of computing the effect of the shift from black and white to color yielded results homogeneous enough to encourage a general statement about the effect of printing in color rather than in black and white.

Suppose, however, that we express the three increases, following the fishing paradigm, as percentage of all readers "still in the pond" who had *not* yet seen the advertisement; we then obtain for the three advertisements the measures given in Table 1-8.

This measure of increase is so similar for the three advertisements that we should be encouraged to accept it as a general estimate of the effect of shifting from black and white to color: color will increase readership by 17 to 19 percent of the readers not yet reached by the black-and-white advertisement.

The logic behind this method of percent computation is this: the higher the starting point in readership, the more difficult it will be to increase readership by whatever means. Hence, to

TABLE 1-8

*Readership Increase from Black and White to Color Measured
in Terms of the Readers Not Yet Reached*

Advertisement	(a) Percent of Readers *Not* Reached by the Black and White Advertisement	(b) Percentage Point Increase If in Color	Gain Expressed as (b) of (a) %
A	100 − 42 = 58	+10	17
B	100 − 23 = 77	+14	18
C	100 − 16 = 84	+16	19
Average increase			18

compute the percent increase on the basis of the potential number of readers not yet reached is likely to provide the best approximation.

SUMMARY

Translation of numbers into percentages facilitates the perception of certain numerical relations. Often these relations are suggestive in the sense that they hint at the causes that account for the differences or changes in the observed percent figures. That potential function of percent figures imposes a special burden on the analyst. By equalizing different base numbers to the 100 percent level, we emphasize certain relations and move others to the periphery of the reader's attention. Percent figures thus function as a causal model, albeit a very rough one. R. A. Fisher, one of the founders of modern statistics, called this "discounting a priori the effects of concomitant variates."[2] The percent comparison will be justified to the extent to which such a priori reasoning proves to be correct.

[2] R. A. Fisher, *The Design of Experiments* (London: Oliver & Boyd, 1942), p. 164.

2

Presentation Problems

From what we have seen to be the purpose of percent figures, certain rules can be derived about their proper presentation. Violation of these rules may impede the simplification process designed to increase the transparency of numerical relations.[1]

NUMBERS AND PERCENTS

There are two kinds of statistical tables. One is a repository of all sorts of numbers, its virtue is its completeness. The other kind, the one we are concerned with here, tries to help the reader perceive its numerical structure at a glance. Too many numbers on such a table endanger its legibility, and percent distributions are particularly vulnerable to such crowding. The absolute numbers should be kept to a minimum; and if they have no meaning of their own, as in sampling surveys, they should be omitted altogether. Only the base numbers need be retained, especially in sampling operations, where they are one of the determinants of the sampling error.

If for some reason it is desirable to retain the absolute numbers alongside the percentages, one should distinguish the two typographically, either by putting one of them into parenthesis,

%	Number
18	(123)

using italics,

18	*123*

---

[1] Several of the principles outlined here have a distinguished if rudimentary ancestor in H. Higgs and G. V. Yule (eds.), *Statistics by the late Sir Robert Giffen written about the years 1898–1900*, (New York: Macmillan, 1913).

or printing the percent figure in boldface type,

18   123

The use of different colors opens additional possibilities.

Columns of absolute numbers, even if typographically distinguished, impede the comparison of percent figures in the horizontal direction, because reading them requires the reader's concentration on every other number. This difficulty can be removed by printing percents and numbers in the following fashion:

%	Number	%	Number
18		25	
	(123)		(212)
82		75	
	(560)		(636)
100		100	
	(683)		(848)

If these devices prove too cumbersome or take too much space, one can always print two tables—one containing the percents; the other, the numbers.

PERCENTS RUNNING OVER 100

For good reasons, we expect percent columns to add up to 100. More often than not, they do; but if they add up to more than 100, the reader should be informed of the reason; namely, that for some or all of the units in a particular column, there was more than one entry or answer. This happens typically when multiple reasons or multiple preferences are given in response to survey questions. A study of the reasons that motivated juries to decide criminal cases differently from the way the presiding judge would have decided them provides a good example. In about 20 percent of all jury trials, the jury disagrees with the judge. The search for the reasons for those disagreements yielded Table 2-2.[2]

Showing at the bottom of the table the average number of items per unit warns the reader that the percentages add up to more than 100—in this case to 164. The average number, of course, may not always suffice; at times, the specific combinations of the multiple reasons or preferences may be important.

SUBTOTALS

Subtotals, unless set off typographically, tend to confuse. The subtotal classification and the corresponding percent number can be extended at each end of the line. In this way, the subtotals as well as the detailed numbers can be read downward without obstruction, as in Table 2-1.

TABLE 2-1
Totals and Subtotals

	Percent
Subtotal	56
xxxxxxxx	24
xxxxxxxx	20
xxxxxxxx	12
Subtotal	44
xxxxxxxx	33
xxxxxxxx	11
Total	100
(Number of cases)	(555)

DECIMALS

Undue or misguided concern for mathematical precision leads to another frequent obstruction to the clarity of tables. Percent figures, like other measures, can be written with any number of

[2] Adapted from H. Kalven Jr. and H. Zeisel, *The American Jury* (Chicago: The University of Chicago Press, 1966), p. 111.

TABLE 2-2

Reasons for the Disagreement of Juries with the Judge

	Percent of Cases
Different evaluation of the evidence	79
Sentiments about the law	50
Sentiments about the defendant	22
Superiority of counsel	8
Facts only the judge knew, not the jury	5
Number of cases in which judge and juror disagreed	100%
Total trials with judge and jury disagreement	(787)
Average number of reasons per trial	1.64

significant decimals: 170 of 450 is 37.777 percent, 37.78 percent, or, rounded off still further, 38 percent.

It may seem that the more accurately a percent figure is computed and presented, the better it will serve its purpose. But with each added decimal, the percent figure loses something of its simplicity and hence of its readability. If the computation of decimals is carried to extremes, the percent figures will be more cumbersome to read than were the original numbers. Decimals tend to defeat the principal purpose of percent figures and should therefore be used judiciously.

Consider the following set of figures:

(*a*) Number	97	129	292
(*b*) Base (=100%)	(352)	(306)	(344)
Percent (*a*) of (*b*)	27.55	42.14	84.88

The decimals do not facilitate the reading of these percent figures. By rounding them off and retaining only the base, we increase their clarity.

Percent	28	42	85
Base	(352)	(306)	(344)

There are situations, however, where the desire for simplicity must yield to accuracy. Decimals may be important because a large base may make even a small difference statistically significant, as in the following example:

Percent	11.5	11.9	12.4
Base	(9,367)	(10,072)	(10,031)

The same differences based on sample sizes of not more than a few hundred cases would not be significant. Retaining such decimals might carry the spurious impression that they differ significantly when all of them should in fact be rounded off and read 12, 12, 12.

Decimals should also be retained in cases where a repeat survey is planned, the results of which will be compared with the first one. Since one cannot know in advance just how large or small the future differences will be, the starting point for the comparison should be retained with precision.

The general rule to be derived from all these experiences is not very precise but is clear in its directive: unless decimals serve a special purpose, they should be omitted. Omitting them will increase clarity and eliminate a spurious appearance of accuracy.

PERCENT, PER MILLE, PER 100,000

In some situations, the problem underlying the use of decimals can be solved only by changing the percent (per hundred) ratio to a per mille or even a higher base. Suppose that one wants to compare the suicide rates of different countries. In percentage terms, such a table would look confusing, as illustrated by Table 2-3.

The zeros in front of each number tend to make the numbers appear more alike than they actually are. The remedy is simple: instead of expressing suicide rates in per 100, express them in terms of per 100,000 population (Table 2-4), which is the universal convention. Thus, all significant numbers are to the left of the

TABLE 2-3

Suicides Rates in Various Countries per 100 Population

West Berlin	0.0395	Sweden	0.0169
Hungary	0.0249	United States	0.0108
Austria	0.0224	Netherlands	0.0066
Finland	0.0221	Mexico	0.0019
Japan	0.0173	Ireland	0.0018

Source: *Demographic Yearbook, 1965* (New York: United Nations), Table 44, p. 762.

decimal point and one decimal is retained because the differences between countries are often small and their changes over time are even smaller.

At the other extreme, percentages that run considerably over 100 fail in their purpose. To state that the profit of company X increased by 2,703 percent makes a formidable figure; but the increase is easier perceived if we are informed that profits are now 28 times as large as they were before.

Percentages are simply a special form of ratios; they should not have more digits than are needed and should be free of zeros at either end.

TABLE 2-4

Frequency of Suicides in Various Countries

(Per 100,000 population)

West Berlin	39.5	Sweden	16.9
Hungary	24.9	United States	10.8
Austria	22.4	Netherlands	6.6
Finland	22.1	Mexico	1.9
Japan	27.3	Ireland	1.8

SPECIAL RATIOS

The base figures are sometimes chosen not because they provide optimal simplification but because they constitute what one might call a natural base, as in Table 2-5.

TABLE 2-5
*Availability of Department Stores
in Three Cities*

	Number of Inhabitants per Store
Edinburgh	10,000
Manchester	16,000
Coventry	22,000

After I. R. Vesselo, *How to Read Statistics* (Princeton, N.J.: Van Nostrand, 1965).

The figures are large and have many zeros, but they have a direct meaning both for the store owners and for the consumers they serve; that the figures are rounded off helps to see their significance.

There are other situations in which a special ratio fits better into the natural frame of thinking than do percent figures. The ratio of officers to enlisted men in an army can be shown as a percentage of the whole, but ratios convey a more vivid picture. Table 2-6 presents the same relation in three different ways.

The percent ratio (*c*) is clearly the least useful way of presenting the relation. Of the two others, (*a*) is perhaps preferable, because it reflects the natural combination.

Some ratios, such as the sex composition of a population, are standardized in a special way. Instead of showing the proportion, it has become the custom—albeit somewhat sexist—to show the number of men per 100 women.

The ratio method would seem to have the advantage here,

TABLE 2-6
Ratio of Officers to Enlisted Men

(*a*) Number of soldiers per officer	250
(*b*) Number of officers per 1,000 soldiers	4
(*c*) Percent of officers of all members of the armed forces	0.398

TABLE 2-7

*Sex Ratio (Men per 100 Women) in Alaska, Utah,
and Washington, D.C., 1960**

| | Percent | | Men per |
	Male	Female	100 Women
Alaska	57	43	132
Utah	50	50	100
Washington, D.C.	47	53	88

* These territories were selected because they represented balance and two extremes.

because there is something like a "normal" ratio: a balanced population—one female for every male. Moreover, in an imbalanced population, the ratio provides a more meaningful measure of the deviation. If a woman is offered a job in Alaska and incidentally would like to know what her chances of finding a husband there would be, "57 percent male" is of little help. To learn that there are 132 males for every 100 females is a more direct measure of her chances.

WHEN CAPTIONS NEED EXPLAINING

In most tables it is not difficult to explain what each number refers to, because both column and line descriptions are, as a rule, simple references to age, sex, income, and other self-explanatory categories. Sometimes, however, the categories will be more complex because of the subject matter of the table. In an Austrian study designed to discover the psychological effects of long-lasting unemployment, it was found to be advisable to distinguish the psychological state of the unemployed families in terms of four categories in descending order with respect to their proximity to the psychological breaking point (see Table 2-8).[3]

[3] Marie Jahoda, Paul F. Lazarsfeld, and Hans Zeisel, *Marienthal, The Sociographie of an Unemployed Community* (New York: Aldine-Atherton, 1972). First German edition, Hirzel, Leipzig, 1933.

TABLE 2-8
The Psychological State of the
Unemployed Families

	Percent
Unbroken	16
Resigned	48
In despair	11
Apathetic	25
	100

Although the four labels were distinctively suggestive, they gave only a general idea of the rich reality behind them. The labels were therefore expanded in two stages—first, by a summarizing explanatory description such as this one for the "resigned" family:

No plans, no relation to the future, no hopes, extreme reduction of all needs beyond the bare necessities, yet at the same time maintenance of the household, care of the children, and an overall feeling of relative well-being.

The second expansion of the label went into even greater detail by describing some of the individual families that exemplified the category.

THE PERCENTAGE CHAIN

A peculiar problem of computing and presenting percent figures arises with respect to figures that form a "decision tree" with "branches" and "twigs." Table 2-9, which itemizes the disposition of criminal defendants whom a grand jury indicted, provides an example of such a chain of decisions. This table cannot be comprehended at a glance, but given the fact that it provides a great deal of information, it can be read with relative ease. The indictment may be dismissed or upheld (line 1). If it is upheld, the defendant must plead guilty or go to trial (lines 2 and 3). The

TABLE 2-9

Disposition of Indicted Defendants in an American Court*

Percent Not Convicted		Percent Convicted	Line

Indictments
(100%)

19.8 dismissed ←——————— ▼ ——→ 80.2 upheld

		go to		plead		(1)
		21.1 trial		guilty	59.1	(2)
		(26% ←——— ▼ ——→ 74%)				(3)
	judge		jury			(4)
14.6	trial		trial 6.5			
	(69% ←— ▼ ———→ 31%)					(5)
5.1 acquitted			convicted	9.5		(6)
(35% ←———— ▼ ————→ 65%)						(7)
2.1 acquitted			convicted	4.4		(8)
(32% ←——————— ▼ ——→ 68%)						(9)

27.0 Total dismissed or acquitted   Total convicted   73.0   (10)

* The data from Dallin H. Oaks and Warren Lehmann, *A Criminal Justice System and the Indigent, A Study of Chicago and Cook County* (Chicago: The University of Chicago Press, 1967.)

first of each pair of lines gives the percent distribution based on the total number of indictments; the percent figures in parentheses are based on the number of decisions represented by the preceding line; by definition, they always add up to 100. If the defendant decides to go to trial, he or she can choose (line 3) between trial before a jury or trial before a judge without jury (lines 4 and 5). In either trial, the defendant may be convicted or acquitted (lines 6 and 7 as well as 8 and 9). Line 10, the sum of all preceding dispositions, gives the overall result of the process: 27 percent of the defendants in this group are dismissed or acquitted; 75 percent are convicted.

OF 100 PEOPLE

A colleague of mine once remarked that large numbers as well as percent figures have a tendency to dehumanize societal relations. I could never see the point until I came across a demographic thumb-nail sketch of our planet that did not use large numbers and, at least nominally, even percent figures:

If the world were a global village of 100 people, 70 of them would be unable to read, and only one would have a college education. Over 50 would be suffering from malnutrition, and over 80 would live in what we call substandard housing. Of the 100 residents in that global village, six would be Americans. These six would have half the village's entire income; and the other 94 would exist on the other half.[4]

GRAPHS

In this book I have used graphs sparingly. Almost all numbers are presented in tables—in a form that makes them easy to read. That is as it should be, because instructing the reader about the proper way to present statistical tables is one of the aims of this book.

More often than not, however, statistical tables are presented to an audience that is neither particularly able nor eager to read them. In such a case, transforming the numbers into graphs is a step that further clarifies numerical relations. This step may also transform the reading of statistics from a chore to a pleasure.

There is a rapidly growing literature on how to and how not to make graphs. The trend has been accelerated by the advent of the computer.

What follows aims at supporting that trend through examples of three kinds of charts: those that build additional information into a basically simple structure; charts that clarify at a glance relationships which mere numbers would reveal only to the expert reader; and, finally, charts made by masters.

TABLES AS BAR GRAPHS

As a rule, a statistical table that conveys much information is more difficult to read than is a table that has little information. There are occasions, however, when having additional information in a

---

[4] "Fellowship of Reconciliation," *Fellowship Magazine*, February 1974. The numbers have somewhat but not radically changed during the intervening decade.

table will increase its readability. Consider a table that shows
how our 50 states stand with respect to the death penalty: 13 do
not have it; 26 have it but have not executed anybody during the
past 30 years; 11 states have executed prisoners. Listing the
names of these states, provided each is accorded equal space,
will facilitate reading by the bar chart these lists provide (see
Table 2-10).

A similar opportunity arises when the numbers in a statistical
table denote units which are described in more detail elsewhere.
At times, for instance, it is desirable to establish a link between a
statistical table and the individual cases of which it is composed.
Figure 2-1, taken from a study of criminal law procedures, serves
this dual purpose. It summarizes the relation between the severity
of the sentence imposed on an offender after conviction and his
prior criminal record.

The table presents graphically the respective frequencies by
the number of squares in each cell. In addition, all cases carry
their number so that the analyst can locate them easily if he
desires to study them in detail.

COMPUTATIONS AT A GLANCE

There are certain numbers, such as the weighted average, the
derivation of which is difficult to convey through a statistical table
because they require a series of multiplications or divisions. A
graph, such as that shown in Figure 2-2, can solve such problems
easily.

The statistical table that contained the four arrest rates would
pose a mild puzzle that is clearly solved by the graph: the average
arrest rate for all felonies is as low as it is because the nonviolent
property crimes provide the great bulk of the arrests.

THE "GESTALT"

Certain configurations reveal themselves only if seen in their
totality. Given the opportunity, we can perceive their structure
in an instant. Figure 2-3 compares the development of the

TABLE 2-10

The Death Penalty in the 50 States of the United States

Have the Death Penalty

No Executions since 1976	Had Executions since 1976	Have No Death Penalty
26		
Arizona		
Arkansas		
California		
Colorado		
Connecticut		
Delaware		
Idaho		
Illinois		
Kentucky		
Maryland		
Massachusetts		
Missouri		
Montana		**13**
Nebraska		Alaska
New Hampshire	**11**	Hawaii
New Jersey	Alabama	Iowa
New Mexico	Florida	Kansas
New York	Georgia	Maine
Ohio	Indiana	Michigan
Oklahoma	Louisiana	Minnesota
Pennsylvania	Mississippi	North Dakota
South Carolina	Nevada	Oregon
South Dakota	North Carolina	Rhode Island
Tennessee	Texas	Vermont
Washington	Utah	West Virginia
Wyoming	Virginia	Wisconsin

Source: Compiled by the Legal Reference Fund, NAACP, June 1984.

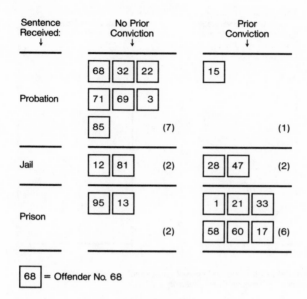

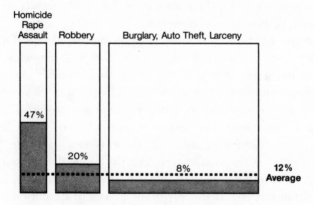

FIGURE 2-1
Relation of Sentence to the Offender's Prior Criminal Record
(The Numbers Identify the Defendant and His Case File)

FIGURE 2-2
Arrest Rate for Felonies Reported to the Police

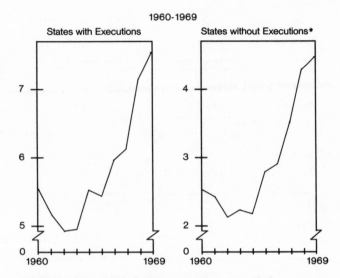

1960-1969

States with Executions   States without Executions*

*Abolition states and 6 states with no executions since 1948.
Source: FBI Uniform Crime Reports

FIGURE 2-3

Homicide Rates, 1960–69, in States With and Without Executions

homicide rate in the states that had executions for murder with the rate in states that either don't have the death penalty or had no executions over the same period.

The graph conveys vividly information that would be difficult to derive from a statistical table. First, that the homicide rate varied considerably during those years; and second, that their development showed an uncanny parallelity. These two perceptions lead to these crucial insights: (1) there are powerful forces in society that move the homicide rate up and down in an outspoken, nonrandom fashion; (2) these forces, to judge from the parallelity of the two curves, are the same throughout the United States, so the deterrent threat of possible execution is not among these forces; (3) the deterrent power of executions, if it exists at all, must be so small as to be practically invisible.[5]

[5] From Hans Zeisel, *The Limits of Law Enforcement* (Chicago: The University of Chicago Press, 1983), p. 61.

ONE PICTURE IS WORTH A THOUSAND NUMBERS

One of my teachers in Vienna was the great Otto Neurath—physicist, sociologist, analytic philosopher, reformer—who passionately fought for pictorial presentation of statistics, as part of a language that would be universally understood. We see its beginning in airports and airplanes all over the globe.

Neurath has greatly enriched the art of statistical graphics by his imaginative application of a basic directive: to transform both words and numbers of a statistical table into easily perceived and easily understood pictures or pictograms as he named them. If the caption of a table involved geographic distribution, a simplified map would provide the pictorial background. The counted objects would be shown so that both their substance and their frequency could be taken in at a glance. Each pictorial symbol would represent a round number of units and their arrangement would facilitate the multiplication leading to their total number. Words and numbers are relegated to a subordinate position. Figures 2-4 and 2-5 are two of his many ingenious graphs.[6]

Figure 2-4 makes its two points with clarity: the skewed shape is the income distribution, along which are the different places occupied by whites and blacks. The symbols effectively remind us that we are dealing not only with dollars but with human beings.

Figure 2-5 depicts a dramatic correlation, albeit one that now belongs to history.

Finally, Figures 2-6 and 2-7 both happen to deal with safety—the one with safety on the highways, the other with the safety of our earth. The law, and we ourselves, talk about things safe and unsafe as if there were a natural dividing line between them. Figure 2-6 was designed to show that safety is rather a matter of degree, which can be increased by appropriate arrangements.

Figure 2-7 allows us to see at a glance a problem we all thought

[6] From Otto Neurath, *Modern Man in the Making* (New York: Alfred Knopf, 1939).

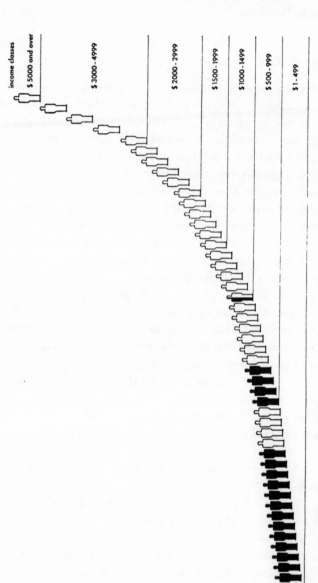

income classes

$ 5000 and over

$ 3000 - 4999

$ 2000 - 2999

$ 1500 - 1999

$ 1000 - 1499

$ 500 - 999

$ 1 - 499

Each symbol represents 2% of families   white: whites   black: Negroes

FIGURE 2-4

Profile of Family Income of Blacks and Whites in Columbia, South Carolina, 1933

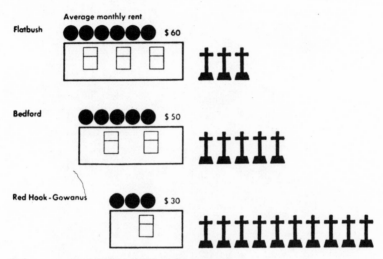

Each cross represents 1 death per 10,000 population per annum

FIGURE 2-5
*Living Conditions and Mortality from Tuberculosis in Brooklyn,
1929–1932*

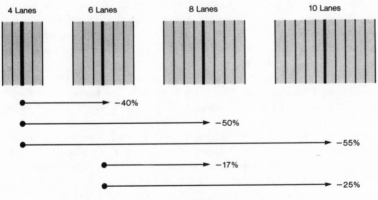

FIGURE 2-6
Accident Reduction Through Adding of Highway Lanes

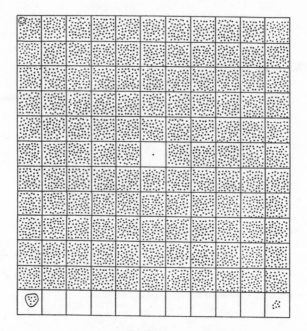

FIGURE 2-7

The Nuclear Weapon of the World

The world's current (1983) firepower in contrast to that of World War II. The dot in the center square represents all the firepower of World War II: 3 megatons. The other dots represent the world's present nuclear weaponry, which equals 6,000 World War II's, or 18,000 megatons. The United States and the Soviets share this firepower with approximately equal destructive capability.

The top left-hand circle, enclosing 9 megatons, represents the weapons on only one Poseidon submarine. This is equal to the firepower of three World War IIs and enough to destroy over 200 of the Soviet's largest cities. We have 31 such subs and 10 similar Polaris subs.

The circle in the lower left-hand square, enclosing 24 megatons, represents one new Trident sub with the firepower of eight World War IIs. *Enough to destroy every major city in the northern hemisphere.*

The Soviets have similar levels of destructive power.

Just two squares on this chart (300 megatons) represent enough firepower to destroy all the large- and medium-size cities in the entire world. (U.S. Senate staff have reviewed this chart and found it to be an accurate representation of the nuclear weapons arsenal.)

to be too complex for us to understand. By visually relating our present state of armament to a few familiar markstones, we instantly comprehend the enormity of our situation. It is probably the most important chart ever made.

SUMMARY

The clarifying function of percent figures is often impeded by their awkward presentation. Percent figures of more than two digits and too many decimal places are difficult to read. At times, special ratios will be superior to ordinary percent figures. If the original numbers accompany the percent figures, they tend to get in one another's way. Typographical devices may help. Transformation into graphic form is always attractive, often essential, and probably the way of the future.

3

In Which Direction Should Percents Be Run?

Cross tabulation, a device that in table form relates two or more factors to each other, presents the analyst with a peculiar problem. In which direction should the percent figures be computed, vertically or horizontally? Computer printouts routinely provide both, which points up the need to choose.

THE CAUSE-AND-EFFECT RULE

The general rule—we will discuss the exceptions later—is this: *Whenever one factor in a cross tabulation can be considered the cause of the other, percents will be most illuminating if they are computed in the direction of the causal factor.*

Consider Table 3-1, which reports the number of men arrested on a felony charge in New York City and the corresponding number of men who had no such arrest. Women were omitted, because in this field of endeavor, they are far from having reached parity and they so far have not aspired to reach it. Women account for less than 10 percent of all felony arrests.

In Tables 3-2 and 3-3, the numbers of Table 3-1 are replaced by their respective percent figures. In Table 3-2 they are run vertically, and in Table 3-3, horizontally.

Table 3-2 points out certain imbalances in the distribution by age of the men arrested and men not arrested, suggesting that the arrest rate is relatively high in the age groups under 29 and relatively low in the age groups beyond that limit. But the exact disproportion can be learned only from Table 3-3, where the percentage of arrested men in each age group is stated irrespective of the size of that age group. From this table we learn that the peak arrest rate (7.1) is in the 16- to 19-year age bracket. It

TABLE 3-1

Annual Number of Men Arrested on a Felony Charge, by Age

(New York City, 1982)

Age Range	Arrested on a Felony Charge	No Such Arrest	Total
10–15	11,959	340,041	352,000
16–19	18,691	245,309	264,000
20–29	25,185	694,815	720,000
30–39	9,900	526,100	536,000
40–49	3,564	388,436	392,000
50+	2,059	1,029,941	1,032,000
Total	71,358	3,224,642	3,296,000

Source: Adapted from Fig. 19 of Hans Zeisel, *The Limits of Law Enforcement* (Chicago: The University of Chicago Press, 1983).

drops to half that size (3.5) in the 20- to 29-year group, and again to one-half (1.8) in the next 10-year bracket and again to one-half (.9) during the next 10 years.

TABLE 3-2

Annual Number of Men Arrested on a Felony Charge, by Age

Age Range	Arrested on a Felony Charge %	No Such Arrest %
10–15	16.7	10.6
16–19	26.2	7.6
20–29	35.3	21.6
30–39	13.9	16.3
40–49	5.0	12.0
50+	2.9	31.9
Total percent	100.0	100.0
(Number)	(71,358)	(3,224,642)

TABLE 3-3

Annual Number of Men Arrested on a Felony Charge, by Age

Age Range	Arrested on a Felony Charge %	No Such Arrest %	Total %	Number (thousands)
10–15	3.4	96.6	100	(352)
16–19	7.1	92.9	100	(264)
20–29	3.5	96.5	100	(720)
30–39	1.8	98.2	100	(536)
40–49	.9	99.1	100	(392)
50+	.2	99.8	100	(1,032)

The reason why one cannot make this comparison easily from Table 3-2 is that the size of its percentages depends in the first place on the size of the respective population segments. If, for instance, there were more younger people, the percentages in the first column would be larger, and the remaining two percentages accordingly smaller. The percentages in Table 3-3 are independent of the respective population sizes.

Thus, the proper way to run the percents for Table 3-1 is in the direction of age, clearly the causal factor in that relationship; in the language of the statistician, age is the independent variable and the arrest rate, the dependent variable.

This rather universal relation between serious crime and age resembles startlingly the age curve discovered by Kinsey for sexual activity; the crime curve, however, ends earlier.

THE AMBIGUITY OF "CAUSE"

As between suicide and age, there is no doubt about which of the two we consider to be the cause and which the effect. Even so, the real cause of suicide might be an incurable disease; and age may appear as "cause" only because older people more often

TABLE 3-4
Remedies for Four Selected Ailments

Ailment	Alka-Seltzer	Anacin	Aspirin	Others	Total
Neuralgia	107	47	198	24	376
Cold	98	41	401	30	570
Upset stomach	302	60	. . .	23	385
Headaches	. . .	242	210	26	478
(Number of reports)	507	390	809	103	1,809

have such diseases than do younger people.[1]

At times we will have no way of knowing which of two factors is the cause of the other, as, for instance, in a table that shows that youngsters who play a musical instrument are more likely to attend concerts. It may be the other way around; youngsters who go more frequently to concerts may be more motivated to learn to play an instrument. Finally, there are relations in which either factor may be considered the cause: the one because it is the natural cause; the other because its properties can be controlled, as in the following example.

Table 3-4 presents the results of a survey in which respondents were asked, "What is your preferred drug remedy for neuralgia, cold, upset stomach, headaches?"[2] Table 3-4 records the answers.

In Table 3-5 the ailment is considered the cause for selecting the respective drug: percents are accordingly computed in the horizontal direction of Table 3-4.

Table 3-5 answers the question, "What is the relative importance

---

[1] More of this problem is discussed in Chapter 9.

[2] I apologize on two grounds for retaining this table from the earlier edition. First, it perpetuates the impression that these remedies are different. The analgesic component in all three is aspirin, a fact that has been kept almost secret. Second, the data are antiquated. An aspirin-free neuralgesic has begun to dominate the market. Again, however, the same chemical substance appears under different names, at very different prices.

TABLE 3-5
Favorite Remedies for Four Ailments

Remedies	Neuralgia %	Cold %	Upset Stomach %	Head- aches %
Alka-Seltzer	28	17	78	. . .
Anacin	13	7	16	50
Aspirin	53	71	. . .	44
Others	6	5	6	6
Total	100	100	100	100
(Number of respondents)	(376)	(570)	(385)	(478)

of these remedies for each ailment?" Of all respondents 53 percent suffering from neuralgia chose aspirin; for patients of the common cold, aspirin is an even more preferred remedy; 71 percent will take it. For upset stomach, 78 percent preferred Alka-Seltzer; and for headaches the choice lay evenly between Anacin and aspirin.

But Table 3-4 can also be looked at from another point of view. One may ask, "To what extent will those suffering from these four indispositions be attracted by the therapeutic promises of each of the remedies?" Table 3-6 gives the answer by computing percents in the other direction.

Table 3-5 allows the manufacturer to estimate potential market share among persons suffering from each of the ailments. Table 3-6 allows the manufacturer to gauge for his product the relative importance of each of the ailments.

DIGGING OUT THE INFORMATION

Table 3-7 is a good example of a presentation that clarifies hardly anything. Even the percent figures obscure rather than illuminate. The failure to distinguish percents and numbers typographically is an added nuisance.

TABLE 3-6

Importance of Ailment for Remedy

	Alka-Seltzer %	Anacin %	Aspirin %	Others %
Neuralgia	21	12	24	23
Cold	19	12	50	29
Upset stomach	60	15	. . .	22
Headache	. . .	62	26	26
Total	100	100	100	100
(Number of reports)	(507)	(390)	(809)	(103)

The percent figures give us the distribution of suicides and suicide attempts by sex and age. Looking first at the actual suicides, we see that their distribution among men and women differs by hardly more than one percentage point. The sex distributions of attempted suicides differ somewhat more, but their age structure is by and large the same. Comparing now actual with attempted suicides, we find that the oldest age group provides a much greater part of the actual suicides than of the attempted ones. The inadequacy of the percent figures given in Table 3-7 is demonstrated best by showing in Table 3-8 the better way of mining the information provided by Table 3-7.

By running percents in the direction of the demographic causes, we learn that male suicide attempts are generally more successful; that the sex difference declines with age, and is most marked among persons under 20; and that it disappears entirely in the oldest age bracket, where suicide attempts are successful at a rate of 91 or 90 out of every 100.

THE PROVISO OF REPRESENTATIVE SAMPLING

The rule about the direction in which percents should be computed must be suspended at times because of statistical limitations of the sample. It might be preferable to compute percents in a

TABLE 3-7

Actual and Attempted Suicides in Japan, 1961

| Age | ACTUAL | | | | ATTEMPTS | | | |
| | Male | | Female | | Male | | Female | |
	Number	Percent	Number	Percent	Number	Percent	Number	Percent
Under 20	1,115	10.4	797	11.0	898	16.7	1,299	23.3
20–40	4,904	45.8	3,202	44.1	3,995	74.6	3,892	69.9
40+	4,687	43.8	3,257	44.9	467	8.7	381	6.8
Total	10,706	100.0	7,256	100.0	5,360	100.0	5,572	100.0

Source: Criminal Statistics, National Police Agency, Tokyo.

TABLE 3-8

Percent of Successful Suicides of All Suicide Attempts by Sex and Age

(Japan 1961)

	Male	Female
Under 20	55	38
20–40	55	45
Over 40	91	90
All ages	67	57

certain direction, but the character of the sample may forbid this.

Consider the following example. During two days of a given week, observations were made at the hosiery counter of a New York department store with respect to the price range of stockings bought and the age of the customers who bought them. Table 3-9 presents the findings in numbers.

The promotion department wanted to learn which price brackets proved most attractive to the various age groups. Price was thus considered as a cause of attracting various age groups, and

TABLE 3-9

Price of Hosiery by Age of Customers*

Age	Price Range, $			
	0.59–0.99	1.00–1.29	1.30–1.79	Total
Up to 34	265	12	130	407
35–49	240	140	208	588
50 and over	32	110	25	167
Total	537	262	363	1,162

* Observations were made on a Thursday and a Saturday, a long time ago when things were cheaper than they are today.

TABLE 3-10

Price of Hosiery and Age of Customers

Age	Price Range, $		
	0.59–0.99	1.00–1.29	1.30–1.79
	%	%	%
Up to 34	49	5	36
35–49	45	53	56
50 and over	6	42	7
Total	100	100	100
(Number of cases)	(537)	(262)	(363)

percentages were accordingly computed in the vertical direction, as shown in Table 3-10.

From Table 3-10 the promotion manager concluded that only 6 percent of the lowest-priced hosiery was bought by women age 50 and over, and that 36 percent of the most expensive stockings were bought by women up to 34, and so forth. He forgot that these customer observations had been made on Thursday and Saturday, the two days on which the age distribution of female customers is not at all typical. The average customer age on these two days is below normal, because it is primarily the younger, working women who shop on these two days: on Thursday because the department stores remain open until late, and on Saturday because it is the working woman's free day. Table 3-10 is therefore misleading. Although Table 3-11 too is derived from Table 3-9, it is not misleading, if one may assume that the young and the older women on Thursdays and Saturdays display the same pattern of hosiery buying as do the young and the older women during the other four days of the week—which has turned out to be the case.

Thus for the most part elderly women buy medium-priced hose;

TABLE 3-11

Price of Hosiery by Age of Customers

Price Range, $	Up to 34 %	35 to 49 %	50 and Over %
0.59–0.99	65	41	19
1.00–1.29	3	24	66
1.30–1.79	32	35	15
Total	100	100	100
(Number of cases)	(407)	(508)	(167)

women in the youngest age group prefer the cheaper stockings; and those in the middle-aged group show preference for the most expensive range. If one knew the age distribution of the female customers throughout the week, one could construct a reasonably correct version of Table 3-10 by weighting the various age groups so that their distribution in the table matches the distribution in the store for the entire week.

The lack of representativeness as obstacle to the cause-and-effect rule is even clearer in the following example. To estimate the relative strength of the two political parties before a certain election, a poll with 2,000 voters was conducted in two states, with the results shown in Table 3-12.

If we compute the percentages in the vertical direction of Table 3-12, Table 3-13 is obtained, which seems to indicate that 42

TABLE 3-12

Party Votes in Two States

State	Republicans	Democrats	Total
A	625	1,375	2,000
B	875	1,125	2,000
Total	1,500	2,500	4,000

TABLE 3-13
Importance of State for Party

	Republicans %	Democrats %
State		
A	42	55
B	58	45
Total	100	100
(Number of interviews)	(1,500)	(2,500)

percent of the votes for the Republicans and 55 percent of the votes for the Democrats come from state A.

A simple consideration shows that this table makes little sense. It is true that 42 percent of the sample votes for the Republican party come from state A. But the 42 percent, and, in fact all four percent figures in Table 3-13, are the result of two components: the relative strength of the parties in each state, and the decision to make an equal number of interviews in each state, although state B has about twice as many voters as does state A. Again, however, by appropriate weighting of the numbers in Table 3-12 a correct, although still not very interesting, Table 3-14 can be derived.

The only sensible way of computing percentages for Table

TABLE 3-14
Importance of Party for State

	State A %	State B %
Republicans	31	44
Democrats	69	56
Total	100	100
(Number of interviews)	(2,000)	(2,000)

3-12 is in the horizontal direction. The cause for the relative strength of the different political parties is assumed to be the state—that is, its particular social and economic structure and the political attitudes of its inhabitants.

The table shows that the Democratic party is the stronger one in both states, but that its superiority is greater in state A.

THE TOTAL COLUMN

It is customary in survey work to have a Total column either at the front or the end of every table. Mathematically, this column is the sum of all breakdowns; and as such, its legitimacy is beyond dispute. But for the reader the Total column has a more general significance. Table 3-15, on the Sunday school attendance of American children, will serve as illustration.

TABLE 3-15
*Sunday School Attendance of American Children
at Different Socioeconomic Levels**

(1935)

	I	II	III	IV	V	Total
Attend regularly	47	43	52	37	32	43
Irregular or not at all	53	57	48	63	68	57
Total	100	100	100	100	100	100
(Number)	(521)	(568)	(461)	(467)	(312)	(2,329)

* From John E. Anderson et al., *The Young Child in the Home* (New York: White House Conference on Child Health and Protection, 1936). The five levels represent: I = professional; II = semiprofessional; III = clerical; IV = farmers; V = semiskilled.

Taking our cue from the heading of the table, we will read the Total column as indicating that 43.3 percent of all American children regularly attend Sunday school. But in this we would be wrong, because the number of interviews at each socioeconomic level does not correspond to the relative frequencies of these

TABLE 3-16
Relative Size of Socioeconomic Groups in the United States

Economic Level	Sample Used in Table 3-15	U.S. Population Census	Difference
I	22	8	+14
II	25	16	+9
III	20	18	+2
IV	20	29	−9
V	13	29	−16
Total	100	100	0

levels in the country's population. Table 3-16 indicates the extent to which the distribution of the sample deviates from that of the actual population.

The upper levels I and II are overrepresented in the sample; the lower levels IV and V are underrepresented. Since Table 3-15 showed Sunday school attendance to be higher in the upper classes, we know that the 43 percent figure in the Total column must be too high. If we weight the attendance percentages of Table 3-15 not by the size of the sample as it is but as it ought to be according to Table 3-16, we obtain 40 percent as the average of all American children regularly attending Sunday school instead of 43 percent.

The maldistribution of the sample endangers the correctness of the Total column only if the breakdowns vary with respect to the crucial variable; if the different socioeconomic levels had not shown differences in Sunday school attendance, the Total column would have been correct in spite of its failure to represent all economic strata in proportion.

SUMMARY

In cross tabulations, percents can be computed in two directions. Normally, they are helpful only if computed in one direction: in the direction of the causal factor. The rule has one constraint: the sample must be representative in that direction. Related to this requirement is the propriety of presenting a Total column. It is justified only if it forms a true sample of the population designated by its heading; otherwise, it must be corrected by appropriate weighting of its components.

4

How to Handle Don't Knows and No Answers

At the bottom of many statistical tables, one finds a category labeled variously "Don't Know," "Not Known," or "No Answer." Because these categories are often small, they seldom receive the attention they deserve. They are made to look like simple failures that may or may not deserve reporting. The problems these categories raise, however, are both intricate and interesting. Their resolution hinges on the exact meaning of these negative responses.

THE LEGITIMATE DON'T KNOWS

The Don't Know answer is not always a failure. Often it is a legitimate answer that must be reported along with the replies of respondents who did know or thought they knew the answer.

Suppose that in the course of some trademark litigation a survey is being conducted to find out whether the trademark of company Y is likely to be confused with the trademark of company X. The response categories of such a survey will be the ones shown in Table 4-1.[1]

The Don't Know answers in this survey are obviously not the result of bad interviewing; they are rather one of the natural answers a respondent may give if asked, "What product or company is identified with this trademark?" Opinion polls often try to ascertain how well the citizenry is informed about certain public issues, and thus may ask, "Do you know the approximate size of the United States defense budget?" or "Would you know the name of the present prime minister of Great Britain?" In all these cases the survey maker knows the correct answer—to what

[1] See Hans Zeisel, "Statistics as Legal Evidence," *International Encyclopedia of Social Sciences* (New York: The Macmillan Company, 1968); and "The Uniqueness of Survey Evidence," *Cornell Law Quarterly*, vol. 45, 1960, p. 322.

TABLE 4-1
Identification of the Trademark of Company Y

	Percent
Identified correctly with company Y	25
Identified incorrectly	
With company X	21
With other companies	3
Did not know to what company it belonged	51
Total, %	100
Number of cases	(498)

company the trademark belongs, the size of the defense budget, and the name of the British prime minister. What the survey maker wants to learn is whether the interviewee knows it too.

Other questions of this sort may be directed at intentions, opinions, and value judgments—at the respondent's views of what ought to be done: "For which party do you intend to vote in the coming election?"; "Is our government, on the whole, pursuing the right policy or the wrong policy in Central America?"

To learn that a voter does not yet know whom he will vote for, or does not know whether the government's policy is correct, must be as important as to learn her views on those issues.

THE FAILURE DON'T KNOWS

The situation is different when the investigator asks a question not in order to test the respondent's knowledge or views but to learn certain facts for which the respondent is the only or best source. The questions the U.S. Census asks are prime examples. If we are asked in which country our grandparents were born or in what year we immigrated into the United States, the Census wants to know the answer; if we answer, "I don't know," or "I don't remember," or do not answer at all, it will be indeed a failure. The interviewer in these situations must go out of her way to obtain the correct answer.

WHAT NOT TO DO WITH THE FAILURE DON'T KNOWS

How is one to treat the Don't Knows that are genuine interviewing failures; is one allowed to retain them as a category in the table just like the legitimate Don't Knows? Let us look at an example. In analyzing the fan mail to a once-distinguished radio program, an effort was made to determine the sex of the writers from their first names. Since in many cases it was impossible to decipher the name, and a few names do not reveal the sex, a fairly large Sex Unknown category was obtained (Table 4-2). The inclusion of that category is improper, because it violates one of the basic rules of setting up statistical tables: categories must be mutually exclusive, with no overlapping, so that any given answer or item fits into only one category.

TABLE 4-2

Sex of Fan-Mail Writers to
"America's Town Meeting"*

		Percent
Sex known		82
Male	54	
Female	28	
Sex unknown		18
Total		100
(Number of letters)		(1,390)

* From a study by Jeanette Sayre at the Bureau of Applied Social Research of Columbia University.

The Sex Unknown label is shorthand for "either male or female," and thus overlaps with the other two categories. As a result, the figures of 54 percent male and 28 percent female in Table 4-3 are misleading. The true figure could lie anywhere between 54 and (54 + 18) 72 for men, and between 28 and (28 + 18) 46 for women, depending on how many men and women

TABLE 4-3

Sex of Fan-Mail Writers by Socioeconomic Status

	Upper Strata %	Lower Strata %
Men	57	52
Women	29	28
Sex unknown	14	20
Total	100	100
(Number of letters)	(450)	(940)

there were among the 18 percent Sex Unknown. This type of consideration is particularly pertinent when different subgroups of the sample show different percentages of the failure category, as in Table 4-3. On the basis of such criteria as paper quality, cleanliness, letterhead, spelling, spacing, punctuation, form of salutation (Table 4-3), the socioeconomic status of the letter writer was estimated. That the reliability of such a determination is low, is irrelevant for the point we are about to make.

Looking at the first line of the table and comparing 57 percent with 52 percent, the casual reader may assume that there were more male fan-mail writers from the upper economic strata than from the lower. It is easy to overlook the point that the low figure of 52 percent is an artifact: it is low because the percentage of Sex Unknown is higher in that group.

Thus, to leave the failure Don't Knows or No Answers in the body of the table is not a good idea, especially when that category is fairly large.

WHAT TO DO WITH THE FAILURE DON'T KNOWS

The crucial question is whether these failures to answer are random events or whether they are associated with certain types of respondents, or have even a meaning of their own.

TABLE 4-4
Sex of Fan-Mail Writers

	Percent
Men	66
Women	34
Total	100%
Number of letters	(1,140)
Number of letters "sex unknown"	(250)
Total letters	(1,390)

If the proportion of the response failures is negligible, one may often simply forego further inquiry and resolve the difficulty by a note such as this:

In the following series of tables, the total number of cases varies slightly because of the exclusion of a small number of No Answer cases, a number that varies slightly from table to table.

If the number of failures to answer is not negligible, their treatment will depend on the result of our inquiry about their randomness. If there is no evidence to the contrary, one may assume that the failures occurred randomly. That assumption implies that if the No Answers had answered, they would be distributed in approximately the same proportions as the known answers are distributed. The No Answers are then best excluded from the body of the table and their frequency noted at the bottom, as in Table 4-4.

If tests revealed that the proportion of No Answers differed significantly among the various subgroups, one might try to reestimate the true total proportions of the known answers by assuming that the random hypothesis is valid only within each subgroup. One would then compute the percent distribution of the known answers for each breakdown and weight the percentages by the fraction which this breakdown (including the No Answers) constitutes of the total sample. By adding the weighted breakdown totals, a new Total column will emerge which is likely to be closer to the true distribution than was the original, natural distribution. As always, however, such weighting is unlikely to effect major changes.

REDUCING THE NUMBER OF DON'T KNOWS

At best, a large number of response failures are only an economic loss; at worst they impede correct interpretation. The proper remedy is prevention; their number should be kept to a minimum. A review of some of the situations that cause Don't Knows to be frequent will suggest some preventive remedies.

Don't Knows will creep in whenever it becomes difficult for a respondent to come up with the correct answer, assuming that the respondent has no reason to hide the truth. In such situations the investigator should try to help.

A Don't Know reply, for instance, to the question, "What brand of flour do you use?" can be changed by the interviewer's suggestion that the respondent look into her pantry.

Sometimes an answer will be withheld because the question, as asked, either is too broad or aims at a somewhat difficult answer. In a pilot survey on shoe buying, the interviewee was asked, "In what kind of store did you buy your shoes?" The question brought raised eyebrows and many Don't Knows from the interviewees, who had a hard time guessing just what types of stores the questioner had in mind. To avoid this difficulty, the question was changed to:

In what kind of store did you buy these shoes: a department store? a chain store? or another type of store?

The change reduced the proportion of Don't Knows only slightly. More detailed questioning revealed that the bulk of these

Don't Knows stemmed from the respondents' inability to state whether the particular store was a chain or "another type" of store; purchase in a department store was always clearly remembered. The number of Don't Knows was drastically reduced when the respondents were given just two choices—department or other stores.

ASKING FOR NUMBERS

A frequent source of too many Don't Knows are questions that begin with "How many" or "How often" when an interviewee is being asked for a numerical response. Interviewees in such a situation who know the answer approximately but not precisely may prefer to answer, "I don't know," rather than to err. There are two solutions to such a problem. The insertion of "approximately" into the question will encourage unsure respondents. The preferable solution is to offer frequency ranges, such as "up to 5" "6 to 15" etc.—sufficiently broad (but not too broad) so that the overwhelming majority of respondents will not hesitate to answer.

The loss in accuracy through such a concession is usually more apparent than real, because seemingly accurate answers often betray their imprecision by bunching around the numbers 5, 10, 20 etc.—a sign that the respondents can answer only within an error range of 5 or 10 points.

INDEFINITE NUMERALS

At times, it will be altogether a mistake to ask for numbers, because the respondents will know the answers only in terms of indefinite numerals. The question, "How high are the heels on your shoes?" yields a great many Don't Knows. The solution here is to ask the question in terms of three height brackets as defined by their vernacular headings:

Would you say that the heels on your shoes are high, medium, or low?

TABLE 4-5

Interpretation of Indefinite Numerals

Number of Monthly Movie Visits	Number of Persons Interpreting			
	Rarely	A Few Times	Fre-quently	Often
0	11			
1	12	1		
2	7	2		
3		17		
4		6	2	
5		4	1	
6			10	2
7			7	2
8			9	18
9			1	3
10 and more				5
Total	30	30	30	30

Indefinite numerals should be used only if their meaning in the context of the question is fairly standardized, as it is in the case of shoe heels. If such standardized terms are not available, a problem of translation arises. In response to a question that begins with "How often" or "How many," the investigator will at times obtain such answers as "rarely" or "often" or "occasionally." If half of the respondents answer in terms of numbers and the other half in terms of such indefinite numerals, the analyst will have difficulties categorizing the data.

In a survey about the frequency of movie going, respondents who had used such indefinite numerals were asked to translate them into number ranges; Table 4-5 provides the ensuing translation.

The boxes around the numbers indicate the most appropriate range suggested by these indefinite numerals. "Rarely" and "a few times" can be safely separated from one another; "frequently" and "often," although different, overlap too much to be distinguished.[2]

DON'T KNOWS WITH A SPECIAL MEANING

Similarly, in a mail survey on candy buying (Table 4-6) the question, "What kind of candy box do you prefer for home use?— for gifts?" yielded a high percentage of No Answers.

TABLE 4-6
Type of Candy Box Preferred

	For Home Use %	For Gifts %
Plain box	55	19
Decorated cardboard	6	28
Silk, satin	3	21
Metal	9	16
No answer	27	16
Total	100	100

That the percentage of No Answers is almost twice as high among the buyers for home use as it is among those who bought candy for gifts suggests that it may be sensible to rename the category No Preference and leave the answers where they are.

Automobile owners when asked, "Have you used your car this year more, less, or about as much as last year?" gave the answers shown in Table 4-7.

The unusually high proportion of Don't Knows and the nature of the question made it improbable that the failures came randomly from among the other three groups. A more compelling interpre-

---

[2] This effort may be seen in the larger framework of "quantifying everyday language," a concern voiced by Frederick Mosteller in a letter to *Science* vol. 192, p. 5.

TABLE 4-7

Use of Car in Current Year
Compared with the Preceding Year

Used the Car	Percent
More	25
Less	14
About the same	20
Don't know	41
Total	100

tation was that the car owners who did not know whether they had driven more or less this year than last might have given that answer because the difference, in whichever direction, was too small to be remembered with sufficient precision. To test this interpretation, a slightly more detailed checklist, given to a comparable sample of respondents, yielded Table 4-8.

The interviewees who answered "Maybe a little more . . . or less" clearly corresponded to the bulk of the Don't Knows in the comparable sample of Table 4-7. The additional choice on the checklist sharply curtailed the number of Don't Knows.

TABLE 4-8

Use of Car in Current Year
Compared with Preceding Year

(Redefined)

Used the Car	Percent
More	22
Less	16
About the same	25
Maybe a little more, maybe a little less—cannot say exactly	30
Don't know	7
Total	100

FACILITATING RECALL

There is still another typical failure to respond to a numerical question. When housewives are asked a question such as, "How many cans of kitchen cleanser did you buy during the last month?", quite a few answer, "I don't know." Their number can be reduced by breaking the question down into two parts:

About how often did you buy cleanser during the last month?
How many cans of cleanser do you usually buy at one time?

This question form guides the respondent's memory and eliminates the need for mental calculation. It has, moreover, the advantage of providing additional data: We now have information on the frequency of purchase *and* the average size of each purchase.

REDUCING THE NUMBER OF LEGITIMATE DON'T KNOWS

The failure Don't Knows are always unwelcome. At times, however, the legitimate Don't Knows may also constitute an undesirable answer, namely, if it is not quite clear precisely *what* the respondent does not know.

During World War II, the Office of War Information surveyed public opinion about the conduct of the war. Since the survey topics were complex, a considerable number of Don't Knows were expected during the first round of questioning. The interviewers, therefore, were instructed to follow up with the question, "Why do you say you do not know?" The answers to this follow-up question fell into the following pattern:

1. *General lack of exposure.* I've had so little time to read about those things. My radio hasn't been working. Have not had a chance to read the news. Can't read so good. Don't read

much. Don't listen to it. Haven't heard much about it.
2. *Information for decision is not available to the public.* If I knew the situation I would know how to answer. Lack of information. People are in no condition to answer. We don't know what is best to make a decision. Don't know enough about this war.
3. *Can't make up my mind.* Have weighed both sides, but can't decide. Have not been able to make up my mind. There is so much to consider. I've thought about it, but can't make up my mind. It takes a lot of thinking to decide. I think one thing one time and then read something which changes my mind so I don't know.
4. *No reason given.* Don't know. I wouldn't know. Couldn't answer intelligently. Hard to answer.

In category 2, some of the respondents went even further and specified the information they thought they needed to make up their mind. In such situations, an effort should be made to determine whether the Don't Know response reflects primarily a lack of information or the inability to form an opinion.

STATISTICAL LIE DETECTION

Criminologists claim that the polygraph is able to distinguish false from true answers, by measuring heartbeat, respiration, and perspiration. Occasionally, statistical analysis can perform a similar function.

During World War II, American housewives were interviewed in a mail survey about their rice-purchasing habits. At the end of the questionnaire, they were asked to identify in which of four brackets their family's income fell. About 12 percent of these housewives did not answer the income question and it was not immediately clear whether the cause was randomly distributed negligence or a more specific reason. The analysis given in Table

TABLE 4-9

Housewives Buying Rice in Bulk or Branded, by Income (1939)

	Income				
	Under $1,000 %	$1,000– $1,999 %	$2,000– $2,999 %	Over $3,000 %	Not Stated %
Brand only	32	43	49	56	55
Bulk only	49	34	25	15	16
Both	19	23	26	29	29
Total	100	100	100	100	100
(Number of cases)	(237)	(715)	(364)	(266)	(212)

4-9 of the purchasing pattern in relation to income suggested the solution.

The higher the income, the greater the proportion of housewives who bought branded rice; the lower the income, the more frequently the cheaper bulk rice was bought. The interesting point is that the housewives who did not state their income exhibited almost the same buying pattern as did the group in the over $3,000 bracket. The similarity suggested a reluctance of many housewives to reveal their high-income bracket, a reluctance that became understandable when we learned that the housewives were offered a small reward for their cooperation. The point was corroborated by another table from that survey (Table 4-10).

Again, the group that did not state its income was most similar to the highest income group.

A more complicated evasive pattern emerged from the analysis of a survey of family relations. Teenagers were asked whether their parents had punished them corporally during their early childhood years. A rather high percentage of these youngsters answered, "I don't remember." The high percentage by itself did not necessarily suggest a hidden motive, but the nature of the topic raised suspicion. The standard procedure recommended for

TABLE 4-10
*Average Price Paid for Branded Rice
by Various Income Groups*

Income, $	Cents
Under 1,000	9
1,000–1,999	10
2,000–3,000	13
Over 3,000	14
Not stated	15

such situations was applied to see whether the frequency of the memory failure was related to some significant characteristic of these children. Table 4-11 revealed such a relationship.

The teenagers who had been corporally punished were less likely to have confidence in their parents (45 versus 50 percent); and the children who claimed they did not remember whether or not their parents had beaten them showed the lowest level of

TABLE 4-11
Corporal Punishment and Confidence in Parents

	Children Say They		
	Had Not Been Punished %	Had Been Punished %	Do Not Remember Whether They Had Been Punished or Not %
In parents	50	45	34
In other persons	32	42	50
Cannot state	18	13	16
Total	100	100	100
Number of cases	(282)	(412)	(165)

confidence in their parents (34 percent). The pattern suggested that Do Not Remember was a conscious or unconscious coverup for the resented and perhaps repressed experience. Subsequent interviews with these children confirmed the suspicion.

CENSUS ELIMINATES 207,000 DON'T KNOWS

At one time the population tables produced by the U.S. Bureau of the Census contained a category "age unknown." In the 1910 Census, for instance, we find Table 4-12.

Although that Unknown category was small enough, it constituted both a nuisance and an impropriety: a nuisance because in all tables based on age breakdowns an extra column had to be tabulated, checked, and printed; an impropriety because it was an overlapping category.

For the straight tabulation reproduced in Table 4-12, the assumption that these 0.2 percent persons of age unknown came randomly from all age brackets, and the corresponding redistribution might have been an acceptable correction. But in a cross

TABLE 4-12
United States Age Distribution, 1910

Age	Percent	
Under 5	11.6	
5–9	10.6	
10–14	9.9	
⋮	⋮	
70–74	1.2	
75–79	0.7	
80–84	0.3	
85 and over	0.2	
Unknown	0.2	(actually 0.18 = 169,055)
Total	100.0	

tabulation of age and marital status, for instance, we would then encounter some married or widowed persons of age "below 5." In fact, the Census Bureau had no difficulty in finding out that the Age Unknowns were not random failures but occurred more frequently in certain age ranges. A disproportionately high number were infants under 1 year of age whose exact age in terms of months was not indicated. Also among certain types of residents, e.g., lodgers and hotel guests, there was a higher frequency of Age Unknown.

In the Census of 1940[3] the Bureau therefore set to work to estimate the individual age of each of the 207,211 persons whose age was unknown. Essentially, this was done by locating other age-related information found on his or her Census questionnaire, such as marital status, school attendance, employment status, and age of other members in the family. A fairly accurate estimate could be made for school children if their grade was known; the age of married people was estimated from the age of the spouse on the basis of correlation tables. Where the age could be ascertained only within a certain range, care was taken to distribute these estimates randomly over the total range so as to avoid artificial clusters. An experimental comparison of 4,000 age estimates against the actual age discovered by a second effort revealed that approximately 45 percent were estimated correctly or within a year of the true age. Only 19 percent were in error by more than 5 years.

SUMMARY

The Don't Knows and No Answers at the bottom of statistical tables require attention and often special treatment, depending on whether they are a legitimate result or a failure of data

---

[3] C. Edward Deming, *The Elimination of Unknown Ages in the 1940 Population Census* (U.S. Department of Commerce, January 1942).

gathering. If these failures occur randomly, they can be safely eliminated. If they occur more frequently in one subgroup than in another, different remedies are indicated. The failure Don't Knows pose special problems if they represent wilful evasion of an unpleasant answer.

5

Tables of More than
Two Dimensions

THE PROBLEM OF REDUCTION

A two-dimensional table—one that tabulates one frequency distribution against the other—normally offers no difficulty of presentation. Tables of three or more dimensions, if presented in the traditional manner, lose in clarity.

Table 5-1 is such a three-dimensional table, from a survey of public attitudes toward the proposed late evening opening of a department store. These attitudes presented as they were affected by the frequency with which the respondents shop at night and by the frequency with which they shop at that department store. Table 5-1 is hard to read even though the absolute numbers, except for the base numbers, have been removed.

If one wants to learn, for instance, how the attitude toward night opening varies with the frequency of shopping at the store, one must compare every fifth number in the vertical direction, which makes for awkward reading.

How is one to resolve such difficulty? The solution can be found with the help of geometric logic. We must get rid of one dimension by reducing the several percentages of a column to one single number, which in the language of geometry means reducing a line to a point.

It is generally advisable to reduce the dependent, or effect, variable, which in the present case can be done easily because it consists of only two numbers. Table 5-2 demonstrates this feasibility by using as example the cell in the lower-right-hand corner of Table 5-1, which reports the attitude for all 1,221 respondents irrespective of their shopping habits.

The information conveyed in this table can be written without

TABLE 5-1

Public Attitudes toward Keeping XX Store Open after 5:30 p.m. by
Frequency of Night Shopping and Frequency of Shopping at XX

Frequency of Shopping at XX	Attitude toward Night Opening	Frequency of Night Shopping			
		Twice a Month or More %	Less Often %	Never %	Total %
Frequently	Favorable	90	58	20	45
	Unfavorable	10	42	80	55
	Total	100	100	100	100
	(Number)	(115)	(105)	(74)	(294)
Occasionally	Favorable	85	51	11	42
	Unfavorable	15	49	89	58
	Total	100	100	100	100
	(Number)	(105)	(103)	(171)	(379)
Never	Favorable	41	24	9	23
	Unfavorable	59	76	91	77
	Total	100	100	100	100
	(Number)	(74)	(80)	(97)	(251)
Total	Favorable	76	48	15	40
	Unfavorable	24	52	85	60
	Total	100	100	100	100
	(Number)	(294)	(359)	(568)	(1,221)

loss in one figure by stating that 40 percent of all respondents
favor night opening. Few readers will have difficulties with seeing
automatically the complementary figure, the 60 percent who do
not favor it. If all 16 columns of Table 5-1 are reduced accordingly,
Table 5-3 is obtained, a three-dimensional table that offers all the
optical advantages of a two-dimensional one. In this table each

TABLE 5-2

Attitude Toward Night Opening

(All respondents)

	Percent
Favor it	40
Do not favor it	60
Total	100
(Number of cases)	(1,221)

figure represents the percent of respondents who, in the particular subgroup, favor night opening. With the exception of the base numbers, this relatively small table contains the same information as does Table 5-1. Unlike that table, it tells its story clearly.

In Table 5-3 all important relations can be easily perceived. The first figure, for instance, indicates that of the people who shop at least twice a month at night and shop frequently at XX, 90 percent favor night opening. Both frequency of shopping at XX and frequency of night shopping affect the attitude toward

TABLE 5-3

Public Attitudes toward Keeping XX Store Open after 5:30 p.m. by Frequency of Night Shopping and Frequency of Shopping at XX*

Frequency of Shopping at XX	Frequency of Night Shopping			
	Twice a Month or More	Less Often	Never	Total
Frequently	90	58	20	45
Occasionally	85	51	11	42
Never	41	24	9	23
Total	76	48	15	40

* The figures indicate the percent of respondents in each group who favor night opening.

TABLE 5-4
Schooling by Sex and Income

	Men		Women	
	High Income %	Low Income %	High Income %	Low Income %
More than high school	50	20	40	10
High school or less	50	80	60	90
Total	100	100	100	100

night opening of that store. As one might expect, of the two, frequency of night shopping is the more important factor: The Total row at the bottom drops from 76 to 15 percent, while the Total column at the right moves only from 45 to 23 percent.

THE PRINCIPLE ILLUSTRATED

A graphic presentation of the reduction method will make the critical relations clear. Consider the following three-dimensional relations between sex, income, and level of schooling shown in Table 5-4.

If the four columns of Table 5-4 are rearranged spatially in the form of a cube, Figure 5-1 is obtained. The bird's-eye view of this cube offers the square fourfold Table 5-5 that contains all relevant data.

The complementing percentages, those having merely high school education or less (represented in the cube by the upper, empty part of the column), are the differences between each of the recorded percentages and 100.

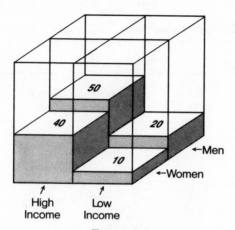

FIGURE 5-1

Percent Having More Than High School Education by Sex and Income

TABLE 5-5

*Percent in Each Group with More
Than High-School Education*

	High Income	Low Income
Men	50	20
Women	40	10

MAKING A DICHOTOMY

What is one to do if the distribution one wants to reduce to a single number is not a natural dichotomy? The answer is: make one anyway. Consider Table 5-6, from a survey of cereal eating.

Here one could combine, for instance, the first three groups and with some loss in detail arrive at the dichotomy of those who do eat cereal and those who never eat it. Or one could set out

TABLE 5-6

Frequency of Eating Cold Cereal, by Sex and Community Size

	Urban		Rural	
	Men %	Women %	Men %	Women %
Daily	27	35	39	31
Frequently	18	22	27	20
Occasionally	6	11	14	10
Never	49	32	20	39
Total	100	100	100	100
(Number of				
cases)	(199)	(201)	(100)	(99)

the daily cereal eaters, the first line, against the sum of the remaining three lines; finally, one could combine the first two lines (those who eat cereal daily or frequently) and contrast them with the third and fourth, those who eat it never or at best

TABLE 5-7

Three Measures of Eating Cold Cereal

(Percent in each group)

Eat Cold Cereal	Urban				Rural			
	Men		Women		Men		Women	
Daily	27		35		39		31	
(Rank)		(4)		(2)		(1)		(3)
Daily or at least								
frequently	45		57		66		51	
(Rank)		(4)		(2)		(1)		(3)
At least								
occasionally	51		68		80		61	
(Rank)		(4)		(2)		(1)		(3)

occasionally. All three methods of reducing the distribution to a dichotomy are presented in Table 5-7. Although the level of the percentages changes, their relative standing varies little and their rank order remains the same. Hence any of these artificial dichotomies will do.

AN AVERAGE REPRESENTING THE COLUMN

If the columns consist of a variable with precise numbers, as in Table 5-8, an average, usually the mean, will conveniently represent the column.

TABLE 5-8

Frequency of Magazine Reading in Two Cities, by Economic Status

(Percent of people reading the indicated number of magazines)

| Number of Magazines Read | City M | | City N | |
	Upper Economic Strata %	Lower Economic Strata %	Upper Economic Strata %	Lower Economic Strata %
0	25	40	16	30
1	23	36	29	42
2	39	18	41	25
3	9	6	10	1
4	4	. . .	3	2
5	1	. . .	1	. . .
Total	100	100	100	100
(Number of persons)	(1,199)	(1,792)	(1,001)	(2,101)

Differences in magazine reading between the two cities and the two economic groups cannot be easily discerned; there are too many figures. From the average number of magazines read

TABLE 5-9

*Average Number of Magazines Read per Person
in Two Cities, by Economic Status*

	Upper Economic Strata	Lower Economic Strata
City M	1.49	0.90
City N	1.58	1.03

per person in each of the four groups (Table 5-9), the interrelations emerge with clarity.

Magazine reading in the upper economic strata is over 50 percent more frequent than it is in the lower strata, and in each group, city N reads slightly more than does city M.

The use of an average to represent a distribution is particularly indicated if the categories of the distribution do not form a natural continuum because they were collected in a form designed to aid the respondent's memory. In the survey of rice consumption already mentioned, housewives were asked how frequently they served rice in various forms. To facilitate their answer, they were given a checklist:

_____ Two or three times a week
_____ Once a week
_____ Two or three times a month
_____ Once a month or less often

To develop average frequencies of consumption, the categories had to be transformed to a common denominator—consumption frequency per month. By some averaging and allowing $4\frac{1}{3}$ weeks per month, the surveyors found that the following weights—17.5, 4.3, 2.5, and 0.5—suggest themselves for the four categories. In this way, the average frequency per month of serving rice can be computed for any group of consumers.

RANK ORDERS

Sometimes, all we have of a distribution is its rank order. Which television channel do you listen to most? Which comes next? Or which brand of bread do you buy (or sell) most frequently? Which next?

To translate such rank-order information into an average requires either more data or some assumptions. One might consider arbitrary weights—4 for the first rank, 3 for the second, and so forth. But how is one to know whether 10, 5, 2, 1 might not be more appropriate weights? Sometimes it is possible to estimate the proper weights from available data, as in the following example.

Table 5-10 presents the answers of 100 grocers in a southern American town to the survey question, "What brands of bread do you sell? State them in order of the approximate dollar volume they represent."

TABLE 5-10

Number of Grocers Reporting Various Bread Brands as the Leading Brand, Second, Third, or Fourth

Brands	Leading Brand	Second Place	Third Place	Fourth Place
A	52	29	13	1
B	28	25	22	12
C	13	11	25	38
D	5	12	15	13
E	1	15	14	15
All others	1	8	11	14
Total	100	100	100	100

The grocers themselves were asked to provide the weights for each rank. They were asked, "Approximately what percentage of your bread sales does your best selling brand account for? and your second brand? etc." The values the grocers gave for each

rank did not differ too widely, and so the surveyor was allowed
to average these rank percentages as follows:

Rank	Percent of Total Sales
1	40
2	31
3	17
4	12
Total	100

By weighting the respective numbers in Table 5-10 by these
percent shares, the surveyor could transform the rank held by
any one brand into its approximate percent share of bread sales,
as shown in Table 5-11.

TABLE 5-11
Market Share of Brands of Bread

Bread Brands	Approximate Percent Share of Market
A	32
B	25
C	17
D	10
E	9
Others	7
Total	100

Occasionally the peculiar structure of a rank-order pattern will
provide an even simpler solution to the transformation problem.
In the early days of broadcasting research, before more precise
listening figures were developed, a surprisingly elegant device
came serendipitously to light for measuring broadcast audiences.
In every county of the United States a representative sample of
households were interviewed by the research department of the

National Broadcasting Company to record all radio stations to which they listened regularly, and also the one station they listened to most often. In some 50-odd cities it was possible to compare the response to these questions with direct measurements of the audience shares that fell to the various broadcast stations in the area. Table 5-12 reveals an interesting relation between the two measurements, that is between the station's share of total listening time and the station's percent share in response to the question, "To which one station do you listen most often?" The proportion of homes that recorded a particular station as Listened to Most turned out to be almost identical with that station's share of the total audience.

TABLE 5-12
*Station Shares of Daytime Radio Audience
in New York City*

	Station "Listened to Most" %	Independently Established Station Share %
WCBS	24	25
WNBC	17	18
WJZ	12	13
WOR	14	14
WNEW	10	10
Other stations	23	20
Total	100	100

In this situation, the answers to a single question provided a shortcut to a reasonably accurate share transformation.

REDUCING A TRICHOTOMY

Survey makers at times ask their respondents to judge a situation in terms of a triple checklist that allows a positive, a negative, and an indifferent answer, such as More, Less, or About the same,

TABLE 5-13

Income Changes during the First Year of World War II, 1941–1942

(Percents in each line representing one occupational group 100%)

	Percent Greater	Percent Same	Percent Smaller	Percent "Greater" Minus Percent "Smaller"
Major executives	29	53	18	(+11)
Minor executives (including professionals)	40	52	8	(+32)
All other professional workers	42	49	9	(+33)
Salespersons	28	56	16	(+12)
Clerks (sales and office)	41	49	10	(+31)
Skilled manual workers	57	36	7	(+50)
Unskilled workers	45	48	7	(+38)
Farmers	39	52	9	(+30)
Not gainfully employed	16	65	19	(−3)
Armed forces	34	22	44	(−10)
Total all groups	40	49	11	+29

Source: "The Impact of War on American Families," *Life*, April 17, 1943, p. 12. Last column provided by the author.

or Like, Dislike, or Indifferent. Table 5-13 is an example. It records the answers to the question, "How does this year's income of yours compare with last year's: Is it greater or smaller or did it remain the same?"

The temptation to contract such a three-number distribution into a single number is great, especially if the numbers are needed for subsequent cross analysis. The simplest contraction will be the mathematical net balance: (percent greater) minus (percent less), with the middle category receiving a weight of zero. This net-balance contraction forms the last column in Table 5-13. This solution is reasonable so long as the middle link is indeed indifferent or inert. It hides, however, differences such as the following:

Distribution A	Distribution B
+20%	+50%
=70	=10
−10	−40
100%	100%
(20 − 10) = +10	(50 − 40) = +10

Both distributions yield the same net balance, +10; but in one the indifferent middle group comprises 70 percent; in the other, only 10 percent.[1]

A FOUR-DIMENSIONAL TABLE

In Sam Stouffer's monumental study of the American armed forces during World War II, there is a table that shows the preference of soldiers for northern or southern training camps, depending on whether they themselves were in one or the other, on whether they came originally from the south or the north, and, on whether they were white or black.[2]

Table 5-14 reproduces that table, relating three causal (independent) factors—regional origin, race, and region of present camp—to one effect (dependent) factor—regional camp preference.

[1] For a discussion of these and other interrelations, see Paul F. Lazarsfeld and W. S. Robinson, "Some Properties of the Trichotomy 'Like,' 'No Opinion,' 'Dislike,' " *Sociometry*, vol. 3, 1940, p. 151.
[2] Sam Stouffer et al., *The American Soldier* (Princeton, N.J.: Princeton University Press, 1949), vol. 1, p. 554.

TABLE 5-14
Soldiers' Preference for Southern or Northern Camp

| | Percent Preferring Camp | | | Number of Soldiers (=100%) |
	South	Indeter- mined	North	
Northern men now in:				
Northern camp				
Blacks	7	18	75	(516)
Whites	11	24	65	(1,470)
Southern camp				
Blacks	18	19	63	(1,390)
Whites	28	24	48	(1,821)
Southern men now in:				
Northern camp				
Blacks	31	25	44	(871)
Whites	49	22	29	(360)
Southern camp				
Blacks	63	23	14	(2,718)
Whites	76	16	8	(1,134)

It is difficult to perceive from Table 5-14 the interrelations among the four factors. However, if expressions of regional preferences are reduced to one figure, as is done in Table 5-15 by subtracting the percent preference for a southern camp from the percent preference for the north, reading the data is easier. This contraction is the more acceptable, since the "middle-group" percentages vary narrowly between 16 and 25 percent.

With the help of this simplified measure of regional preference, Tables 5-16 to 5-18 clarify the respective influences of regional origin, ethnic background, and location of the campsite. Note that all four tables (Tables 5-15 to 5-18) are identical in content; each contains the same eight figures developed in Table 5-14. They differ merely in the form of their arrangement, which is designed

TABLE 5-15
*Soldiers' Preference for Northern
or Southern Camp*

	Preference Score for Northern Camp[*]
Northern men now in:	
Northern camp	
Blacks	+68
Whites	+54
Southern camp	
Blacks	+45
Whites	+20
Southern men now in:	
Northern camp	
Blacks	+13
Whites	−20
Southern camp	
Blacks	−49
Whites	−68

[*] A negative score indicates a preference balance for a southern camp.

to clarify their respective analytic points. Table 5-16 explores how black and white soldiers differ in their preferences under otherwise identical conditions.

Blacks consistently preferred the northern camp more than did whites, irrespective of whether they were originally northerners or southerners, or whether they were in a northern or southern camp; the average difference in preference is 23 points. The difference between blacks and whites is greatest for southerners who are in a northern camp (+33) and smallest for northerners in a northern camp (+14). The difference caused by ethnic background is smallest (14 and 19 points) if the soldiers are located in a camp situated in their home region. Altogether, the

TABLE 5-16
Ethnic Background and Regional Preference

	Blacks	Whites	Point Difference
Northerners in:			
Northern camp	68	54	(+14)
Southern camp	45	20	(+25)
All northerners			(+20)
Southerners in:			
Northern camp	13	−20	(+33)
Southern camp	−49	−68	(+19)
All southerners			(+26)
Total average difference			(+23)

range of variations around the average is relatively small, from 14 to 33 points.

Table 5-17 considers the influence of the location of the campsite on the regional preference.

Regional preference is affected by the camp the soldiers are in. If they are in a northern camp, they are more in favor of a northern camp; if they are in a southern camp, they are more likely to prefer a southern camp. The average difference the camp location makes is 42 points, almost twice as large a difference than the one caused by the respondent's ethnic origin.

The influence of the actual experience, however, is smaller for northerners, whose preference index changes by 28 points, than it is for southerners, whose preference index is changed by 55 points. This means that for southerners, the experience of being in the north affects their original preference for the south more than does the experience of being in the south affect the normal northern preference for the north. White northerners who come to the south are more likely to change their preference (+34) than do black northerners if they come to the south (+23). And

TABLE 5-17
Campsite Location and Regional Preference

	Northern Camp	Southern Camp	Difference
Northerners			
Black	68	45	(+23)
White	54	20	(+34)
All northerners			(+28)
Southerners			
Black	13	−49	(+62)
White	−20	−68	(+48)
All southerners			(+55)
Total average difference			(+42)

the regional preference of black southerners is more affected by the experience of living in the north (+62) than is the preference of whites (+48).

Finally, Table 5-18 looks at the influence of the regional origin of the soldiers.

TABLE 5-18
Regional Origin and Regional Preference

	Northerners	Southerners	Difference
In northern camps			
Blacks	68	13	(+55)
Whites	54	−20	(+74)
All in northern camp			(+68)
In southern camps			
Blacks	45	−49	(+93)
Whites	20	−68	(+88)
All in southern camp			(+90)
Total average difference			(+81)

The most powerful influence on regional preference, not un-expectedly, is the regional origin of the soldiers. The average difference caused by this factor is 81 points, almost twice as large as that caused by the camp location (42 points) and almost four times as large as that caused by the ethnic background (23 points). Preference for the northern camp is considerably stronger among soldiers who came from the north than it is among soldiers who came from the south; the average difference is 81 points.

The influence of the home region, however, is dampened if the soldiers are in a northern camp (+68) as compared with their being in a southern camp (+90).[3]

SUMMARY

Only tables with not more than two variables can be presented in their entirety and still be readable. Additional variables lead to problems of presenting three or more dimensions in a two-dimensional space. Such problems can be solved by telescoping one or more columns into a single figure: a percentage, an average, a ratio—any figure that may adequately represent the column. Which one of these compressions will be most adequate depends on the nature of the measure, the distribution of the measure over the column, and the purpose of the table.

---

[3] Probably because of the charmingly complex interplay of influences in this table from *The American Soldier*, ibid., Leo Goodman also used it in Chapter 1 of his book *Analyzing Qualitative Categorical Data* (Boston: Abt Books, 1978).

6

Indices

The preceding chapter discussed several ways of representing a column of numbers by one figure: an index that stands for the column. This chapter will expand the discussion of indices, by considering their variety and functions and some of the problems of their construction.

We will call an index any single figure designed to measure a multifaceted concept. Changes, for instance, in the cost of living are measured by the consumer price index; a community's state of health may be measured by its rate of infant mortality; the beauty of a young lady may be determined by the rank judges accord her in a contest; a student's intelligence is measured by IQ; and so forth.[1] In each of these instances a relatively complex concept is measured by a single figure. The complexity the index tries to summarize may derive from the multitude of its units, as in the consumer price index, or from the many dimensions of a single unit, as in a beauty contest.

When does one need an index and what purpose is served by such a reduction to one number? Isn't it sufficient, one may ask, in a beauty contest, to say lady X has the most beautiful legs and lady Y the most beautiful face? Someone could say that, but what if one wanted to choose the most beautiful lady overall? What need is there for summarizing changes in the price of food, shelter, and transportation in one number? The answers are several. First, the summary figure tells us by how much the real value of our dollar changes. That number is needed, for instance, for adjusting social security payments and even contractual wages in some industries.

---

[1] To measure psychological and aesthetic phenomena was envisioned by Quetelet who wrote in 1835: "Psychological criteria are in this respect . . . not too different from physical properties: one can estimate their magnitude provided that they are in some relationship to the effects which they produce." [*Essai de Physique Sociale* (Bachelier, Paris, 1835) vol. 2, p. 98.]

Indices are constructed by methods of varying durability and complexity. The ranking of participants in a beauty contest is a one-time affair; the consumer price index is designed to last a very long time. And the complexity of an index, as we will see, may vary from a simple average to intricate mathematical structures.

JUDGMENT INDICES

Most indices are based on objective measurements. Some derive from subjective judgments and others, from a combination of judgment and measurement.

The index in a beauty contest may not be more than the average rank accorded to each contestant by the several members of the jury. In the more serious beauty contests of dogs or horses, the judges are provided with precise guidelines on how to go about their business.

Many sporting contests require no indexing because they have only one dimension: whoever runs fastest or jumps highest wins. Some contests, however, such as diving, skating, or ski-jumping, require indexing because they are judged along more than one dimension; ski-jumping, for instance, by the length of the jump and its style.

We will discuss in some detail the diving index, which consists of both objective and judgmental elements. The objective part of the index is the degree of difficulty of the particular dive. There are now 72 basic dives, for which an international committee of experts has established difficulty values that vary between 1.0 for the simplest dive and 3.5 for the most difficult one.

The judgmental dimension of the eventual diving score has to do with the degree of perfection with which the dive was performed. It is established by a group of expert judges, each of whom decides on a grade from 1 to 10, with a grade of 10 standing for perfection. To avoid bias, the two extreme grades

TABLE 6-1
Diving Score for Contestant X

Performance Scores from Three Judges

	Individual Scores (1)	Average (2)	Standardized Grade of Difficulty (3)	Weighted Score (2) × (3) (4)
Forward diving from 1-meter board; hands tucked	8, 9, 7	8.00	1.0	8.00
Back dive from a 5-meter board	5, 7, 7	6.33	1.6	10.13
Reverse 1-1/2 somersault, running from a 10-meter board; pike position	8, 7, 8	7.33	2.2	16.13
Forward 1-1/2 somersault; three twists; from 3-meter board	4, 6, 4	4.67	2.9	13.54
Total				47.80

given by the judges, the lowest and the highest, are eliminated, and the average of the remaining grades is then multiplied by the difficulty score of the dive. If each contestant must perform several dives, the sum of the several scores is the final index score of the particular diver. Table 6-1 shows how such a performance index is computed for one performer in a four-dive contest.

Indices that are based entirely or even partly on judgments have little durability. Beauty queens from different contests cannot be compared without a new contest. Nor, as a rule, can one

compare divers or gymnasts from different contests, unless the
judges are the same in all of them. Only at the extreme can an
inference of superiority be made. When the young Rumanian
gymnast Comenici won a perfect score, which no other gymnast
had ever achieved before, she was acclaimed as the best ever.
But other gymnasts have obtained perfect scores since, so the
ceiling problem continues.

In institutional setups, comparisons over time may be possible.
Students graded by their professors in one year might be compared
with the students in an earlier year, provided that neither the
grading standards nor the professors have changed in the meantime.

A COMPLEX AVERAGE

We now turn to the consumer price index, which is based on
objective data throughout and therefore by and large retains its
comparability over time. Its purpose is to reflect the average
price changes of the "market basket"—the goods and services
that the American household consumes.

There are conceptual difficulties involved with defining "the
American household," inasmuch as almost no two households buy
the same goods and services in exactly the same proportions. The
difficulty has been partially solved by defining the American
household as the average of the more homogenous category of all
wage-earning households rather than that of all households.

Figure 6-1 presents a schematic rather than real computation
on which the definition of the index is based.

The relative share of the various major budget categories—
food, shelter, health, etc.—is determined by studying the expen-
ditures of the families surveyed in the sample. Certain goods and
services are then selected to represent the broad categories and
their subcategories. Thus the price of a chicken and of one pound

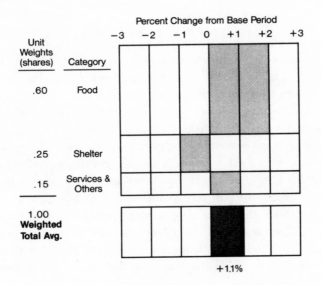

FIGURE 6-1

Schematic Structure of the Consumer Price Index

of chuck steak might represent the price of meat; the price of a bus ride and the price of cleaning a dress or suit may represent services; and so forth. In the end, the prices are combined into subindices for food, shelter, and services, and eventually into the overall index.

The consumer price index is thus the weighted average of many price changes, the weights being determined by the relative share of the budget category which these prices represent.[2]

[2] On the intricacies of these operations see for instance *The Price Statistics of the Federal Government* (a report by the Price Statistics Review Committee, George J. Stigler, Chairman), Joint Economic Committee Hearings, U.S. Congress, Jan. 24, 1961.

INDEX OBJECT AND INDEX FORMULA

The consumer price index, as we have seen, can be fairly easily determined and defined. For other indices, following the path from concept to realization is not that simple. Consider the job of constructing an index of "marital happiness." That task begins by our making explicit what we mean by such happiness or, better still, how we recognize it. Perhaps with the help of a marriage counselor, we might prepare a list of criteria that allow us to distinguish happy from unhappy marriages. In a good marriage, we may decide, the partners spend much time together, converse a lot, observe their wedding anniversaries, and write to one another when separated. In a bad marriage there are quarrels, infidelity, and other unpleasant experiences. These criteria are then formalized and positive and negative values are assigned to them:

Go out together, rather than separately	Almost always	+1
	Half and half	0
	Hardly ever	−1

Observe wedding anniversary	Regularly	+1
	Occasionally	0
	Never	−1

If one criterion, e.g., quarrels, is considered a more significant indicator than any other, or emerges as such from appropriate analysis, a higher value can be assigned to it:

Quarrel	Often	−2
	Seldom	0
	Never	+2

In its most primitive form the happiness index will be the average of such scores.[3]

---

[3] After Ernest W. Burgess and Leonard S. Cottrell, Jr., *Predicting Success or Failure in Marriage* (Englewood Cliffs, N.J.: Prentice-Hall, 1939).

AMBIGUOUS LABELS

Consider an index that claims to measure the prevailing wage
level, a term which, unless more sharply defined, invites abuse.
Table 6-2 compares "wage levels" for two different years and
yields four different answers depending on which wage level we
mean; the numbers are hypothetical.

TABLE 6-2
Comparing 1919 with 1927 "Wage Levels"[*]

	1919 $	1927 $	Change %
Hourly wage rate	1.00	1.25	+25
Daily wage rate	8.00	8.00	. . .
Annual earnings	2080	1870	−10
Annual earnings in 1919 dollars	2080	1680	−19

[*] After Willford I. King, *Index Numbers Elucidated* (New York: Longmans, 1930), p. 29.

The hourly wage rate went up 25 percent. The daily wage rate
remained unchanged because, we are forced to conclude, the
increase in the hourly rate was offset by a reduction in daily
working hours. The annual dollar earnings fell by 10 percent
because the number of working days fell to that extent. Finally
the wage level in real purchasing power fell by an additional 9
percent because of inflation; prices had increased by 10 percent
in the intervening time.

Table 6-3 shows an even more complicated trap caused by the
imprecise definition of wage level.

The daily wage rate for both male and female workers in area
A is 25 percent higher than in area B. Yet the average daily wage
rate for all workers, regardless of their sex, is lower in A than in
B. The apparent reversal derives from the higher proportion of

TABLE 6-3

Comparing "Wage Levels" in Two Areas*

	Labor Force			Average Daily Wage		
Area	Total Workers	Number of Men	Number of Women	For Men $	For Women $	Per Worker (Regardless of Sex) $
A	2,000	1,000	1,000	15.00	5.00	10.00
B	2,000	1,800	200	12.00	4.00	11.20

* Adapted from Franz Zizek, *Statistical Averages* (New York: H. Holt 1913), p. 35.

female workers in area **B** and from the fact that the female wage rate—the year is 1912—is far lower than the male one. All contradictions disappear once the labels are defined with precision.

EFFECT OF "AGING"

One of the more frequently presented comparisons is the difference between persons of different age. On closer scrutiny, the substance of such seemingly simple comparison is complex. The problem is

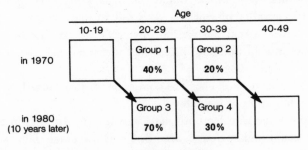

FIGURE 6-2

Proportion Actively Engaged in Sport

(Two Concepts of Aging)

best seen with the help of a hypothetical diagram (Figure 6-2), which compares two age brackets, people in their twenties (20 to 29) and people in their thirties (30 to 39). We want to find out what effect aging has on the proportion of people in each group who actively engage in some sport; we make the comparison twice, in 1970 and 1980, at an interval of 10 years.

The effect of aging can normally be deduced by comparing group 1 with group 2 which, on the average, is 10 years older. The difference, a decline of 20 percentage points, or one-half of all who had been engaged in sport in their twenties, is ascribed to aging.

The precise effect of growing older also depends, however, on the time during which one grows older—often called the generation effect. That effect is best measured by comparing group 1 with 3, and group 2 with 4. The comparison shows that within a decade, involvement in sports has risen from 40 to 70 percent among persons in their twenties and from 20 to 30 percent among persons in their thirties.

Figure 6-2 allows us to see also the combined effect of growing older and of belonging to a different generation. Comparing group 1 with group 4 provides this measure, because group 1 in 1970 has become group 4 10 years later in 1980. That combined effect is a decline by 10 percentage points from 40 to 30 percent. The mathematics is as follows:

Effect of growing older (2 − 1)   −20%
    − Generation effect (4 − 2)   +10%
_____

Combined effect (2 − 1) + (4 − 2) = (4 − 1) = −10%

BASEBALL INDICES

An object is occasionally measured by several indices, each of which measures a different aspect. The three main indices designed to measure the batting performance of baseball players—the batting average, the slugging average, and the number of runs batted in—are examples in point.

Hitting the pitched ball as often and as well as possible, and preventing the other team from making such hits, is the crux of baseball. Hits send players running on their way to one or more of the bases. Subsequent hits by teammates may allow a player to complete the circuit for a run. The number of runs made by each team decides the game.

The batting average (BA) is the best known and most widely used index of a player's batting performance:

$$\text{Batting average} = \frac{\text{number of hits}}{\text{number of times at bat}}$$

The batting average measures a player's batting performance by the relative frequency with which he scores a hit, not necessarily with how often he gets on base. He may get on base not because he hit the ball well but because the opposing team made a fielding error. This event counts as an "at bat" but not a hit, which increases the denominator of the BA index but not its numerator. If the player gets on base because of four bad pitches ("balls") or having been hit by a pitched ball, the event enters neither the numerator nor the denominator, on the theory that it has provided no test of the player's batting ability. For the same reason, the batting average is not affected by a sacrifice hit, whereby the player hits into an out, albeit deep in the field, in order to advance a base runner.

The batting average is not a complete measure of a batter's performance, since it fails to measure two other dimensions of good batting—the quality and timeliness of the player's hits. Quality is measured by the number of bases the player is able to cover before he is forced to stop. If he must stop at first base, he has covered only one-fourth of his track; if he hit the ball deep enough to reach second base, he is half way home, and so forth. The so-called slugging average measures both the frequency and the quality of the hits:

$$\text{Slugging average} = \frac{\text{total number of bases reached}}{\text{times at bat}}$$

TABLE 6-4
Batting and Slugging Average

	Times at Bat	Singles	Doubles	Triples	Home Runs	Batting Average	Slugging Average
		Number of					
Player A	500	100	30	10	10	.300	.400
Player B	500	130	10	10	...	.300	.360
Player C	500	80	10	...	20	.220	.360

Table 6-4 compares batting and slugging average for three hypothetical players.

Players A and B have the same batting average, but A has the higher slugging average; players B and C have the same slugging average, but B has the higher batting average.

The timeliness of a hit is measured by still another batting index—the number of runs batted in (RBIs). It records the number of teammates on one of the bases when the player came to bat who were able to complete their run because of his hit.

It is instructive to consider the relative merits of these three indices. The slugging average is the most comprehensive of the three. When a club considers how much it should pay a player, it will look at that index.

The batting average measures only the frequency but not the quality of hitting. But it has one great advantage over the other two indices: it is easily understood because its mathematics is simple. When a player goes to bat and the loudspeaker advises the spectators that his batting average is "330," everybody understands the meaning: the odds of his having a hit this time at bat is one in three. There is no similarly simple way of translating the slugging average into odds. Moreover, there are many moments in a baseball game when the crucial question is not whether a player will hit well but whether he will hit at all.

The RBI index measures a dimension not covered by either the batting or the slugging average. Not every hit ends up in a completed run; often, at the end of an inning, runners are left on base, an indication of wasted hits. To be able to hit at the right time when runners are on base and could run home is a particularly valuable quality for a baseball player. The RBI index measures that quality, albeit imperfectly. First, it is not related to the frequency at bat, so it grows with the progress of the baseball season. Moreover, it is affected by the order in which a player comes to bat. If he comes after good hitters, who get on base more frequently than do other players, his chances of batting a run in are greater than are those of other players who may be equally good batters.

Even taken together, these three batting indices do not cover the total offensive capacity of a player, because they do not consider his ability to bunt or his ability to run and to steal bases. For such an overall measure, a "scoring index" has been suggested: the probability (measured by past performance) of a player's getting on base multiplied by the probability of his advancing home. This index has the additional advantage of providing the foundation for a team index by averaging the scores of the individual players. And since every player must also perform defensively when the other team is at bat, appropriate defense indices must complete the picture.[4]

Batting indices raise still another problem. Occasionally the question arises about how the legendary baseball heroes of old compare with the present ones, i.e., whether the indices can claim stability over time. That stability is threatened, because a number of changes in the game have at times made it easier or more difficult to hit: changes in the ball; bans on the spit ball; the introduction of better gloves and smoother surfaces, both of which lead to better fielding; changes in the pitching rules; and so forth. To correct and adjust the batting average for such changes,

---

[4] See Earnshaw Cook and Wendell L. Garner, *Percentage Baseball* (Cambridge, Mass.: M.I.T., 1964) chap. 2.

proposals have been made to measure a player's performance by
the ratio between his annual batting average and the average of
all batting averages during that year. A player's career score
would then be the average of all his annual ratios. On such a
measure, Ty Cobb leads the all-time greats with an index ratio of
+102 percent. Next, with an index ratio of +96 percent,[5] is "Say-
It-Ain't-So" Shoeless Jackson, the luckless player whose records
have been expunged because of his involvement in the 1919
Black Sox scandal.

OLYMPIC SCORING

To round out our review of sport indices, we will now look at the
peculiar construction of the measure that decides the winners of
the Olympic decathlon. In each of its ten disciplines, performance
is accurately measured in centimeters or fractions of a second,
dependent on the discipline. But how is one to combine the
measurements of centimeters and seconds, or for that matter the
performance scores in the broad jump with that in the high jump,
both of which are measured in centimeters? And how is one to
decide whether A or B wins if A jumped higher than B and B ran
faster than A?

A fair system must give the same number of points in each
discipline to performances that are of comparable difficulty or
excellence. Table 6-5 lists the equivalent performances in the ten
disciplines that get a score of 800, 900, and 1,000, respectively;
the intermediate scores have been omitted.

In each discipline the performance interval from 800 to 900
points is greater than the interval from 900 to 1,000; in the 100-
meter dash, for instance, the two are 0.4 and 0.3 of a second.
This reflects the fact that a gain of one-tenth of a second becomes
harder as one reaches the top performance level.

[5] See Reed Browning, "These Numbers Don't Lie," *Sports Illustrated*, April
7, 1980, p. 70ff.

TABLE 6-5
Equivalent Scores for the Ten Disciplines

Discipline	Measuring Unit	800 Points	900 Points	1000 Points
100 meters	Seconds	11, 0	10, 6	10, 3
400 meters	Seconds	50, 1	48, 0	46, 0
110-meter hurdle	Seconds	15, 5	14, 5	13, 7
1500 meters	Minutes and seconds	4.02, 0	3.50, 6	3.40, 2
High jump	Centimeters	194	205	217
Broad jump	Centimeters	690	739	790
Pole vault	Centimeters	398	436	478
Shot put	Meters	15, 19	16, 92	18, 75
Discus	Meters	45, 99	51, 58	57, 50
Javelin	Meters	63, 17	71, 81	81, 00

How are these points of equality across disciplines determined?
One begins by collecting for each discipline a sample of performances by competent athletes, defined, for instance, as competitors
in a sufficiently large number of high-level international meetings.
These performances are likely to yield distributions that approximate the normal bell-shaped curve. Such a distribution has two
basic dimensions: the mean and the so-called standard deviation,
or the distance from the mean to the point where the curve
changes inflection, as shown in Figure 6-3.

At the bottom of the figure, the performance levels for the
100-meter dash and the broad jump are put at their appropriate
position in their respective normal distribution. Each of the
juxtaposed performances measures in the two disciplines are
equivalents, i.e., they should and do receive the same number of
points, because both performances are at the same point of
"difficulty" in their respective distribution.

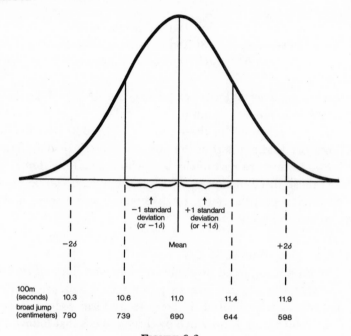

FIGURE 6-3

*The Standard Measures of the Normal Distribution for the
100m Dash and the Broad Jump*

INTERRELATED PERCENTAGES

The following example, from the field of readership research,
presents a set of fairly simple indices that if properly interrelated
cover a broad conceptual framework.[6] From three basic sets of
data, fractions are developed that measure important editorial
aspects of the items offered in the printed media. These data
were developed to help the editors assess the quality of their
pieces and learn the reasons why some of them fared better than
others. A sample of 1,000 readers of the magazine were asked
the following questions for each item in the particular issue:

[6] Originally developed for the now defunct *This Week* magazine. (Matilda
White and Hans Zeisel, "Reading Indices," *Journal of Marketing*, October 1941,
pp. 103–111.)

	(A) Did you see this item?
If yes to *A*:	(B) Did you start reading it?
If yes to *B*:	(C) Did you finish reading it?

For each item, a set of basic percentages was obtained, as shown by way of example in Table 6-6.

There is overlapping in these answers. The 200 who finished reading story MM are part of the 800 who started reading it; and these 800 in turn are part of the 900 who noticed the item.

From these three measures, four reading indices can be developed, which report the item's relative failure and success, and to some extent the cause of it:

Complete readership: ratio C/T

The proportion of all magazine readers who finished reading the item. This is the overall performance measure. It does not tell us what caused the item to perform well or poorly. That information is provided by three indices that follow. For story MM this index was .20.

Attention value: ratio A/T

The proportion of readers who noticed the item when paging through the magazine or, more accurately, who remembered having seen it. This index might be taken as a measure of all factors that account for attracting the reader's initial attention, such as position in the magazine, size of the title, layout, and size and colors of illustration if any. For story MM this index was .90.

Attractiveness of subject matter: ratio B/A

The proportion of readers who started reading the item of those who had noticed it. It is a rough measure of the appeal of the subject matter as suggested by title, illustration, and other factors. For story MM this index was .89.

Holding power: ratio C/B

The proportion of readers who finished reading the item from among those who started reading it. It is a measure of the

TABLE 6-6
Readership Data for Three Stories

	Number of Readers		
	Story MM	Story NN	Story OO
(A) Saw the item	900	300	800
(B) Started reading it	800	300	200
(C) Finished reading it	200	300	100
(T) Total	1,000	1,000	1,000

item's holding power, built up by such factors as attractiveness of content, development of plot, relative length, difficulty of style. For story MM this index was .25.

All four indices may vary between 0.0 representing the poorest possible performance and 1.0, the potential optimum. Like all indices, these too reveal their true usefulness if used in comparative context. Table 6-7 translates the data of Table 6-6 into their respective indices.

Measured by their overall success, complete readership, story NN came out best (.30), followed by story MM (.20) and story OO (.10). The three indices that follow explain *why* the total readership was as high or as low as it was. Story MM was seen by almost everybody (.90), and almost everybody who saw it started

TABLE 6-7
Four Reading Indices for Three Stories

	Story MM	Story NN	Story OO
Complete readership (C/T)	.20	.30	.10
Attention value (A/T)	.90	.30	.80
Attractiveness of subject (B/A)	.89	1.00	.25
Holding power (C/B)	.25	1.00	.50

reading it (.89) because it held high promise; for some reasons, however, the story itself did not hold the reader's interest. Index C/B shows the low value of .25, signaling to the editor that this story's weakness was in its text.

Story NN was seen by only a small proportion of the readers (.30). However, everybody who saw it began reading it (1.00), and the development of the story itself lived up to its promise, because the holding-power index also reached top value (1.00); everybody who began to read the item finished reading it.

Story OO was seen by many readers (.80), but only a few of them started reading it (.25). The suggested content must have been unattractive, because only half who started to read the item finished it (.50).

The precise meaning of these indices was then explored further by the interviewers asking what elements make the attention index high or low; what makes a story look attractive or unattractive to the prospective reader; and what keeps or loses a reader, once she begins reading a story?[7]

SOCIOMETRIC INDICES

The indices described here are based on so-called sociometric attitude scores. They describe the structure of a small group such as a class, a Boy Scout troop, or a workshop. Each member of the group is asked to state her or his attitude toward every other member with the help of a simple five-point scale ranging from maximum acceptance $(+1)$ to complete rejection (-1), with a neutral midpoint at 0 and intermediate values at $+1/2$ and $-1/2$. Table 6-8 presents the mutual acceptance scores for a group of seven members identified by roman numerals. The 1 at the crossing point of member III (top) and II (left) expressed full

---

[7] See, for example, Evelyn Perloff, "Prediction of Male Readership of Magazine Articles," *Journal of Applied Psychology*, vol. 32, 1948, pp. 663–674, and a parallel paper on female readership in the same journal, vol. 33, 1949, pp. 175–180.

TABLE 6-8

*Interpersonal Attitudes in a Group of Seven**

Attitudes Received by Member Number	Attitudes Expressed by Member Number							Total Score Received
	I	II	III	IV	V	VI	VII	
I	...	1	0	0	0	1/2	1	$2\frac{1}{2}$
II	1	...	1	1	1/2	1/2	1/2	$4\frac{1}{2}$
III	0	0	...	0	0	1/2	0	1/2
IV	1/2	0	1/2	...	1/2	1	1/2	3
V	1/2	−1	0	1/2	...	1	0	1
VI	0	1/2	−1	−1	0	...	0	$-1\frac{1}{2}$
VII	0	0	−1/2	0	1/2	1/2	...	1/2
Sum of the scores given to	2	1/2	0	1/2	1-1/2	4	2	$10\frac{1}{2}$

* Taken from Leslie D. Zeleney, "Status: Its Measurement and Control in Education," *Sociometry*, 1941, vol. 4, p. 198.

acceptance of member II by member III.

From these scores we can develop an astonishingly large number of indices that describe the variety of relations in this group of seven members.

The first six indices describe the individuals; index 7 characterizes the relation between pairs; index 8 characterizes the group.

Index 1, Mean Score Received

Measures the individual's acceptance by the group. It is obtained by dividing the numbers in the last column of the table, the total score received, by 6. It may vary, as the individual scores do, between +1.0 and −1.0. The indices for each member are: I (.41), II (.75), III (.08), IV (.50), V (.17), VI (−.25), and VII (.08). Member II has the highest acceptance score and member VI the lowest.

Index 2, Average Deviation from the Mean Score Received

The degree of unanimity by which index 1 was attributed to each individual. A zero deviation means that the same score was accorded to the individual by the other six members of the group. The average amount by which the six individual scores deviate from index 1 for each of the seven individuals is: I (.42), II (.25), III (.14), IV (.17), V (.50), VI (.50), VII (.25). The group verdict about II is relatively homogeneous (.14); acceptance of V and VI shows the greatest variation from one member of the group to the other (.50).

Index 3, Mean Score Expressed

Measures the active sociability of the individual—the degree to which she, in turn, accepts the other members in the group; I (.33), II (.08), III (0.0), IV (.08), V (.25), VI (.67), VII (.33). Number VI shows the greatest readiness to accept others (.67); III (0.0), the smallest.

Index 4, Average Deviation from the Mean Score Expressed

The degree to which each individual discriminates in his acceptance of other members; I (.33), II (.43), III (.50), IV (.43), V (.25), VI (.22), VII (.33). Number VI differentiates least in accepting his colleagues (.22); III shows the greatest degree of differentiation (.50).

Index 5, Correlation between the Scores Given to the Other Members and the Scores Received by Those Members

Measures by the Spearman coefficient of rank correlation the degree to which feelings are reciprocated for each member of the group: I (.17), II (.43), III (−.19), IV (−.51), V (.17), VI (.17), VII (−.11). Individual II reciprocates most accurately the acceptance he receives from the group (.43). Number IV scores his colleagues pretty much contrary to the acceptance he receives from them (−.51); and of VII, one might say that the acceptance he accords is independent (−.11), i.e., close to zero, from the one he receives from his colleagues.

Index 6, Correlation between the Scores Expressed for Each Individual and the General Mean Score (Index I) of Acceptance of Each Individual

The degree to which the scores expressed by each individual conform to the general opinion as expressed by the group as a whole.[8] The coefficients are: I (.94), II (.14), III (1.00), IV (.94), V (.69), VI (.71), VII (.83). Number III scored his colleagues exactly as the group as a whole scored them (1.00); II deviated more than any other member in the group (.14). That all the coefficients are positive indicates that there is on the whole agreement among the members.

[8] Here it becomes particularly apparent that in a small group such as this one, the attitude of one individual may have a disproportionate effect on the group score; the effect would be negligible if the group were large.

*Index 7, Mean of Score Given and Score Received between
Any Two Individuals, which Indicates the Affinity between
These Two*

Only one of the $(6 \times 7:2) = 21$ pair relations in the group attained
the highest possible mutual score $(+1.00)$, namely, members I
and II. The poorest relations $(-.25)$ are found for the pairs II and
V, III and VI, and III and VII. No worse scores than $-.25$ appear.

Index 8, Mean of All Scores

Measures the cohesion of the group. With the aid of this index
different groups may be compared, and changes within the group
may be measured over time. The mean of all scores given (and
received) in the group is $(10:42) = .25$. Considering that the
hypothetical maximum score is 1.00 (if all scores expressed full
acceptance), and the minimum score is -1.00 (if all expressed
full rejection), we can say that this is a moderately cohesive
group.

SPEARMAN COEFFICIENT OF RANK CORRELATION

We now embark on what may seem a gratuitous enterprise, the
construction of an index that has long been in existence and can
be found in any elementary statistics text: the Spearman coefficient
of rank correlation, which measures through one figure the
degree of similarity or dissimilarity between two sets of rankings
of the same items. We will use as basis for our construction
exercise the three different rankings of five contestants, as shown
in Table 6-9.

Without much computation it can be seen that judge A is closer
to judge B than he is to judge C, but it is not immediately clear
whether judge B is closer to judge A or to judge C. It is one of
the purposes of such an index to provide a measurement sufficiently

TABLE 6-9
Ranking of Five Contestants by Three Judges

Contestants	Judge A	Judge B	Judge C
K	1	2	3
L	2	1	2
M	3	3	1
N	4	5	5
O	5	4	4

sensitive to distinguish close rankings.

How are we to start the development of a similarity index? We will begin again by defining the two limits our index can reach—complete identity and maximum dissimilarity. In Table 6-10 these two extremes are denoted for a scale of five items, as in our example.

These two pairs of rank orders have a mathematical property that holds true for scales of any size: The sum of the products of each pair of rank orders reaches a maximum for complete similarity (identity) and reaches a minimum for complete dissimilarity. These multiplications and additions are carried out in the third

TABLE 6-10
Rank Correlation of Five Items

Identity			Maximum Dissimilarity		
(a)	(b)	(a) × (b)	(a)	(b)	(a) × (b)
1	1	1	1	5	5
2	2	4	2	4	8
3	3	9	3	3	9
4	4	16	4	2	8
5	5	25	5	1	5
		55			35

and sixth column (a) × (b) of Table 6-10: the sum of the products
for complete similarity is 55; the corresponding sum for maximum
dissimilarity is 35. The difference between these two extreme
values is 20. The product sums of all other possible combinations
of five-item scales lie somewhere between 35 and 55.

These limits, of course, vary with the length of the scale; for a
four-item scale they are 20 and 30; for a six-item scale, 56 and
91, and so forth. Since we would like to build our index formula
so that it can be applied to scales of any size, it is advisable to
construct it so that the maximum value (identity) always yields an
index of +1.0, and complete dissimilarity always an index of −1.0.
The midpoint of that range 45 for the five-item scale, 25 for the
four-item scale, should always equal 0.0, indicating the point of
no correlation.

The rest develops simply. Determine the interval between the
maximum score and the midpoint; for the four-item scale it is
30 − 25 = 5; then divide the index interval between +1.0 and
0.0 into five equal sections; and assign these values to the scores
between 30 and 25 so that 29 becomes +0.8, 28 becomes +0.6,
and so forth. Figure 6-4 shows this development for the four-
item, five-item, and six-item scale.

The index values are the respective values for the Spearman
coefficient of rank correlation. It is, of course, much simpler to
derive them through the Spearman formula, the more elegant
equivalent of our cumbersome but instructive derivation:[9]

$$p = 1 - \frac{6 \sum D^2}{N(N^2 - 1)}$$

Σ (sigma) stands for "the sum of," D for the difference between
the two ranks given to any one item, and N for the number of
ranks, that is, of the number of items in the scale.

There remains the task of applying the mathematics to our

[9] Its mathematical development can be found, for instance, in William G.
Cochran, *Statistical Methods*, (Ames: Iowa State University Press, 1967), p. 194.

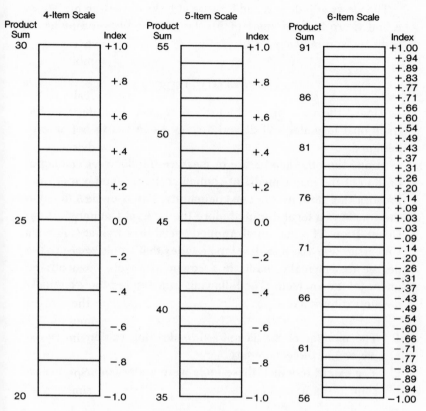

FIGURE 6-4

Development of the Spearman Coefficient of Rank Correlation

example of the five contestants ranked differently by the three
judges. Table 6-11 gives the p values for the three pairs of judges:

TABLE 6-11

Rank Correlation between Three Judges

Judges	Coefficient of Correlation
A and B	+.8
A and C	+.5
B and C	+.6

The views of judges A and B turn out to be closest (+.8); judges A and C are farthest apart (+.5); the relation between judges B and C falls in between (+.6).

CUSTOM-MADE INDICES

At times, the indexing problem is so special that none of the traditional formulas will fit and a special one has to be custom-designed.

At one time the need arose to measure the degree of monopolization of the major media of communications—newspapers and broadcasting stations—in one community. The issue developed in the course of a legal dispute before the Federal Communications Commission. The Bureau of Applied Social Research at Columbia University was asked to develop an index that would measure the degree to which the media in a community were concentrated under joint ownership. The following data entered the eventually developed formula:

1. The number of media units (broadcasting stations or newspapers) in the community.
2. The extent to which these units were under common ownership.

A community, for instance, might have two newspapers and two broadcasting stations, and one newspaper and one of the stations might be under joint ownership. To develop the index formula a number of conditions were set forth, derived partly from the concept of monopolization and partly from certain formal requirements that would make the index clear and practicable.

1. The index would not distinguish between broadcasting stations and newspapers; for purposes of the index each is one medium (M).
2. The index was to vary between 0.0, if there is no joint

ownership between any of the media, and 1.0 if all media are under joint ownership (complete monopoly).

3. The index is to grow as the number of competing media decreases: in a city with four units, M-M, M-M should have a higher index value than M-M, M, M.

4. The number of competing units being equal, the index is to be higher, the greater the inequalities between the competing media: M-M-M, M should produce a higher index value than M-M, M-M.

The formula finally developed[10] was

$$I = \frac{\sqrt{X_1^2 + X_2^2 + \cdots + X_n^2}}{N}$$

N stands for the total number of media in the community. X_1, X_2, ..., X_n stands for the number of media in each joint ownership group; X stands for 2 if two media are under one joint ownership, for 3 if three are under joint ownerships, and so forth; n is the total number of joint ownership groups. Single ownership media are not entered into the numerator; hence, if all units are under separate ownership, the numerator, and therefore the index, has the value of zero.

The formula establishes first the ratio of joint ownership units to the total number of units, but gives the former in terms of the square root of the sum of the squares, to ensure that the constellation M-M-M, M has a higher value than the combination M-M, M-M, although both communities have two competing groups.

The formula assigns the following index values to the five

[10] By Paul Zeisel in cooperation with the author. For the discussion of two sophisticated indices, designed to measure concentration in *any* industry and their validation either through an underlying theory or empirical correlates, see M. O. Finkelstein and R. O. Friedberg, "The Application of an Entropy Theory of Concentration to the Clayton Act," *Yale Law Journal*, 1967, p. 677, and the note there by George Stigler, p. 718.

possibilities of joint ownership of four units: (1) M-M-M-M: $\sqrt{4^2}/4$ = 1.00 (complete monopolization); (2) M-M-M, O: $\sqrt{3^2}/4$ = .75; (3) M-M, M-M: $2^2 + 2^2/4$ = .71; (4) M-M, M, M: $\sqrt{2^2}/4$ = .50; (5) M, M, M, M: $\sqrt{0}/4$ − 0.00 (perfect competition).

SUMMARY

Constructing an index means expressing a more-dimensional object through one figure. If the object itself is measured in its totality, the index will be a simplified contraction of those measurements. Often, however, the index object is defined conceptually and the index formula tries to do justice to the concept. In one sense the question about what an index measures is tautological: it is what the formula indicates. The concept, however, may transcend the immediate formula and include properties that are merely related to the numerical values of the formula.

PART II

THE TOOLS OF
CAUSAL ANALYSIS

Much of the work in the social sciences is still concerned with
the relatively simple task of describing what is and what happens,
because there are still so many social realms we have not yet
observed with any precision. More and more frequently, however, we are trying to find out why things happen and what
effects they have. No other achievement has marked so significantly the progress of the social sciences as has the improvement
of our ability to explain why people behave the way they do,
and to forsee to some extent the effects of our actions.

The first chapter in Part II, Chapter 7, deals with a very
preliminary step toward causal analysis—the refinement of
statistical data by seeing how certain relations vary from one
subgroup of the population to the other: how they differ as
between young and elderly people, or men and women, or
young women and elderly women, and so forth.

Chapter 8 deals with the well-known, if rarely used, tool for
exploring cause-and-effect relations, the randomized controlled
experiment. We know it best from reports on effectiveness
testing in the pharmaceutical field. It allows us to determine
with some precision whether or not a certain treatment produces
an expected effect. In addition to its usefulness as a research
tool, the controlled experiment is important as the conceptual
paradigm that guides also the analysis of quasi-experimental
and nonexperimental data.

Chapter 9 deals with the problems of analyzing differential
treatment data that were *not* generated through randomized

controlled experimentation. Such nonexperimental—also called observational—data are often the administrative by-product of the normal course of events. Quasi-experiment is the name given to experimental designs which potentially share all the features but one of the ideal experiment; for one reason or another they lack random separation of experimental and control units.[1]

Chapter 10, on regression analysis, points up the radically different power of that research tool if it is applied to controlled experimentation, in contrast to its more frequent application to observational data. In this, regression analysis confronts the same problems as do the more elementary analytical techniques.

Chapters 11 and 12 present a very different research tool for the discovery of cause-and-effect relations. Its pursuit goes in the opposite direction of the controlled experiment, not from cause to effect but from the effect backward to its many causes. The method has become known as reason analysis, or colloquially as "the art of asking why," after a pioneering paper by Paul Lazarsfeld that outlined the range and difficulties of that approach.[2] The counterpart to reason analysis is reason assessment: efforts to reconstruct a chain of events, for instance, of an automobile accident, when the network of causes combines reported actions and motives with technical circumstances revealed only through silent clues.

If the process under observation extends over a longer time span, data collection through the normal research channel, the survey, and even surveys repeated at intervals, will often prove unsatisfactory. As a rule, in such situations we want to record the development of each unit over time; of a voting decision, a health history, a purchasing pattern, or a criminal career. For

[1] Its theory and practice was first systematically developed by Thomas D. Cook and Donald T. Campbell, *Quasi-Experimentation* (Chicago: Rand McNally, 1979).

[2] Paul F. Lazarsfeld's pioneering 1935 paper "The Art of Asking Why" is reprinted in D. Katz et al. (eds.), *Public Opinion and Propaganda* (New York: Holt, 1954).

such tasks, repeated interviews or observations over an extended period of time of a more or less permanent group of individuals or other units—now called panel—offer unique analytic advantages. They are discussed in Chapter 13.

Chapter 14 concludes the book with what has become known as triangulation, gains to be derived from coordinating different approaches to the same research goal. Given the imperfection of all research, such mutual support will often make the difference between doubtful and reasonably certain acceptance.

7

The Cross-Tabulation Refines

PURPOSE OF CROSS TABULATING

The starting point of any statistical analysis is the one-dimensional, straightforward table that shows a distribution among several groups and in its simplest form among two groups, as in Table 7-1.[1] From this simplest of all tables, one can learn that at the time of the poll there was a bare majority for the Republican candidate; and if things did not change, the Republican candidate would carry the county. Thus, such a table has descriptive and to some extent predictive value.

TABLE 7-1
Sample Poll in County X before
Presidential Election

Will vote for	Percent
Republican candidate	52
Democratic candidate	48
Total	100
(Number of cases)	(5,160)

Looked at in another way, such a table is merely the starting point for explorations that proceed by dividing the sample into subgroups in order to learn how the voting pattern varies from one population group to the next. This is the function of the cross tabulation.

[1] Tables 7-1 and 7-2 are adaptations from a study of the Willkie–Roosevelt presidential campaign of 1940, as mirrored in a small Ohio town. That pioneering study was later published in Paul F. Lazarsfeld, Hazel Gaudet, and Bernard Berelson, *The People's Choice* (New York: Columbia University Press, 1948). See also Table 7-12.

TABLE 7-2
Preelection Poll in County X by Economic Status

	Economic Status	
	High	Low
Will vote for	%	%
Republican	60	45
Democrat	40	55
Total	100	100
(Number of cases)	(2,604)	(2,556)

If the new two-dimensional distribution differs from the old one-dimensional one, a step has been taken in the process of discovering the factors that determine these proportions. For example, if the table is broken down by the prospective voters' economic status, Table 7-2 is obtained.

The table shows that the proportion of Republican voters is larger among voters from the upper economic brackets and the proportion of Democratic voters is larger in the lower brackets. Thus, generally speaking, economic status is one factor that determines the proportion of Democratic and Republican votes.

Table 7-3 is another example, taken from one of the many studies of automobile accidents.

If we want to find out what factors characterize the people who

TABLE 7-3
Accident Rate of Automobile Drivers

	Percent
Never had an accident while driving	62
Had at least one accident while driving	38
Total	100
(Number of cases)	(14,030)

TABLE 7-4
Accident Rate of Male and Female Drivers

	Men %	Women %
Never had an accident while driving	56	68
Had at least one accident while driving	44	32
Total	100	100
(Number of cases)	(7,080)	(6,950)

have automobile accidents, we may begin by looking for population subgroups which we expect to have different accident rates. Table 7-4, for instance, explores this question with respect to the driver's sex.

The table shows that more male drivers have accidents than do female drivers. By introducing the additional factor, sex, into the analysis, we refine the preliminary result and shed light on the factors that may determine the original distribution.

DIFFERENT TYPES

The procedure can be extended by injecting alternative factors into the tabulation. Such a series of *alternative* breakdowns—by sex, by age, by economic status, and so on—are the prevalent form in which statistical surveys are presented. Yet, as the following paragraphs show, the results of such alternative cross tabulations are often unsatisfactory and occasionally even misleading. The correct procedure is to introduce each additional factor not as an alternative to but something simultaneous with the other factors, so that all possible interrelations become visible.

The simultaneous introduction of an additional factor may produce any of the following effects:

1. It may *refine* the results of the simple cross tabulation.

2. It may fail to refine the results but may reveal an *independent* effect of a third factor.
3. It may *explain* the results of the simple cross tabulation by *confirming* the original interpretation.
4. It may explain the results by *revealing* the original interpretation as *spurious*.

The remaining part of this chapter discusses the first two possibilities, the refining function of the cross tabulation. Chapter 8 will discuss the cross tabulation's power to explain.

THE ADDITIONAL FACTOR REFINES

Through cross tabulation, users of a certain breakfast food were found to be more frequent among people below 40 years of age than among older people (Table 7-5).

The investigator thought of sex as an additional factor that might influence the consumption of XX breakfast food. The proper way of introducing this new factor is shown in Table 7-6. To simplify the table, the complementing percentages of those persons who do *not* use XX were omitted.

This table presents the relation between age and use of XX separately for two different conditions, for men and for women.

TABLE 7-5
Consumption of Breakfast Food XX, by Age

	Persons	
	Below 40 %	40 and Over %
Eat XX	28	20
Don't eat XX	72	80
Total	(1,224)	(952)

TABLE 7-6
Use of Breakfast Food XX, by Sex and Age

	Men		Women	
	Below 40	40 and Over	Below 40	40 and Over
Eat XX	36%	23%	20%	17%
(Number of cases)	(619)	(480)	(605)	(472)

Table 7-5 showed that a relation existed between age and the consumption of XX. Table 7-6 now refines this knowledge by showing how this age relation differs for the two sexes: age differentiates more sharply among men (36 vs. 23 percent) than among women (20 vs. 17 percent). Figure 7-1 shows how the percentages in Table 7-5 are related to those in Table 7-6. Moreover, by exchanging the order of columns two and three in Table 7-6, this figure emphasizes a different aspect of Table 7-6: instead of considering the age differences by sex, it focuses on the sex difference by age.

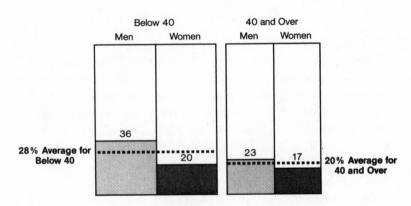

FIGURE 7-1
Use of Breakfast Food XX by Age and Sex

TABLE 7-7

Listening to Classical Music, by Age*

	Below 40	40 and Over
Listen to classical music	64%	64%
(Number of cases)	(603)	(676)

* Adapted from Paul F. Lazarsfeld, *Radio and the Printed Page* (New York: Duell, Sloan & Pearce, 1940), p. 98.

In Figure 7-1, the height of each bar represents 100 percent of the respondents in the particular subgroup; the width of each bar indicates the relative number of persons in each group. The dotted lines represent the weighted average of the users of XX irrespective of their sex: 28 percent among the younger people, 20 percent among the older ones. The solid lines show that in each age bracket there are more XX users among men than among women; the sex difference moreover is greater among the young people than among the older ones: 36 vs. 20 percent as against 23 vs. 17 percent.

CORRELATIONS NEAR ZERO

Situations where the original correlation is zero or near zero are of special interest, because the introduction of a third factor may reveal a submerged correlation. Table 7-7, for instance, shows the frequency of listening to classical music by age.

Contrary perhaps to expectation, there is no correlation between age and listening to classical music. However, when education is introduced into the analysis as an additional factor in Table 7-8, a more complex picture emerges. The various relations are easily seen in Figure 7-2.

The introduction of education reveals that there is in fact a correlation between age and listening to classical music. College-educated people listen more often to classical music when they

TABLE 7-8
Listening to Classical Music, by Age and Education

	Below 40	40 and Over
College	69%	79%
(Number of cases)	(224)	(251)
Below college	61%	56%
(Number of cases)	(379)	(425)

are older: 79 vs. 69 percent. It is the other way around with people on a lower educational level; they listen more to classical music when they are young: 56 vs. 61 percent. If people are grouped by age, regardless of their level of education, these two tendencies cancel one another, reducing the overall correlation to zero.

A particularly interesting example of such a misleading non-correlation emerged from an experiment designed to test the effectiveness of a headache remedy.

The manufacturer of analgesic A was running short of ingredient

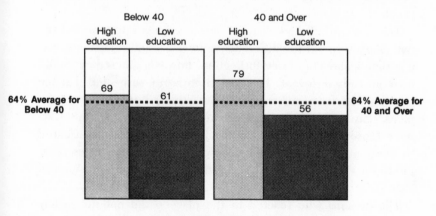

FIGURE 7-2
Listening to Classical Music, by Age and Education

TABLE 7-9
Effectiveness of Three Pills

	Percent Reporting Relief
A	84
A lacking x	80
Placebo	52

Source: E. M. Jellinek, "Clinical Tests on Comparative Effectiveness of Analgesic Drugs," *Biometric Bulletin of the American Statistical Association,* October 1946, pp. 87–91.

x that went into its making. In order to find out whether the absence of x made the analgesic less effective, 200 subjects suffering from infrequent headaches were treated in three successive two-week periods on a rotating basis with three products: the proper drug A, drug A lacking ingredient x, and a placebo, an inactive pill that had merely the appearance of a drug. The success of these three treatments was measured in terms of the percentage of subjects who reported relief from the headache (Table 7-9).

The inactive pill had clearly a lower success rate than did the two analgesics; but the difference between A and A lacking x was not significant. On closer inspection, however, ingredient x did turn out to be relevant. The analyst correctly suspected that the patients who failed to react to the inactive pill would have been more sensitive and therefore more relevant test persons than were those who professed that their headaches had been cured by the placebo. He therefore computed the success rates separately for these two groups and found the reaction presented in Table 7-10.

The persons who reacted to the placebo did not notice any difference between A and A lacking x. But for the subjects who

TABLE 7-10

Proportion of Four Groups of Headache Sufferers Reporting Relief

	Claimed Relief from Placebo	Did Not Claim Relief from Placebo
Percent reporting relief from drug A	82	88
Percent reporting relief from drug A lacking x	84	77

did not react to the placebo, the absence of x made a difference of 11 percentage points. The difference was obscured by being mixed up with an insignificant difference in the other direction, among those unreliable test persons who had claimed relief from the inactive placebo.

ADDITIONAL FACTOR REVEALS LIMITING CONDITIONS

The refinement brought about by the third factor sometimes consists of revealing that certain correlations tend to disappear under special conditions, and correspondingly increase in their absence. The suicide statistic for the French part of Switzerland for the years 1881–1890 showed the annual suicide rate for Catholics to be 19.9 per 100,000 population, roughly one-half the Protestant rate of 39.6. When the two denominations were divided by whether they lived in the city or in the country, Table 7-11 was obtained.

In urban areas the difference between the two denominations is relatively minor: 39.9 vs. 37.8 percent. In rural areas, however, the difference is large: 8.8 vs. 41.4 percent. The data in Table 7-12 permit also a second comparison: The Protestant suicide rate is hardly affected by the urban-rural difference, in sharp contrast to the comparable suicide rates among Catholics. Figure 7-3 presents the two comparisons.

TABLE 7-11

Suicide Rate of the French Swiss 1881–1890
by Size of Community and Religion
per 100,000 Population

Area	Catholics	Protestants
Urban	39.9	37.8
Rural	8.8	41.4
Total	19.9	39.6

Source: M. Halbwachs, *Les Causes du Suicide* (Paris: Felix Alcan 1930), p. 282.

ADDITIONAL FACTOR HAS AN INDEPENDENT EFFECT

Sometimes the third factor may have no effect on the original correlation, and hence will fail to refine it, but will have an independent effect on the factor that in the original cross tabulation was considered the *effect*.

If we introduce religion into Table 7-2, the preelection poll result by economic status, we obtain Table 7-12.

On each economic level, Catholics produce less than half as

TABLE 7-12

Percent Voting Republican in County X
by Economic Status and Religion

	High Economic Status		Low Economic Status	
	Percent	Number	Percent	Number
Catholics	27	(547)	19	(538)
Protestants	69	(2057)	52	(2018)
Total*	60	(2604)	45	(2556)

* See Table 7-2.

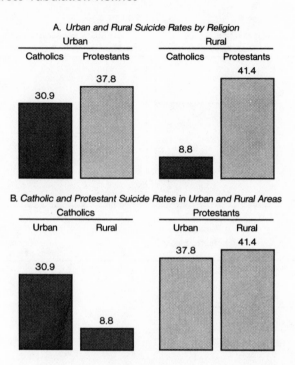

FIGURE 7-3
Suicide Rates of the French Swiss by Size of Community and Religion
(Per 100,000 Population)

many Republican votes as do Protestants (compare vertically); and within each religious group, the higher economic strata produce more Republican votes than do the lower strata (compare horizontally).

Again, it will be helpful to see the relation between these four cells graphically in Figure 7-4. The two graphs make it clear that both factors, economic status and religious denomination, are independently related to the party voted for; hence, the proportion of Republican voters is highest among the well-to-do Protestants and lowest among the poor Catholics.

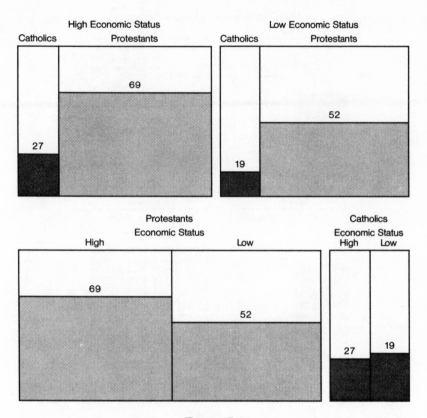

FIGURE 7-4
Vote Intention by Economic Status and Religion

SUMMARY

Cross tabulation, that is, a breakdown of a distribution into subgroups, is the most common device of survey analysis. This chapter discusses a preliminary function of cross tabulation, namely, the setting off of differences in distributions by revealing the subgroups in which certain measures reach extreme values. In which population stratum is the Republican vote most concentrated? Who is most likely to listen to classical music? Which population segments have the lowest suicide rates? This refinement operation lays the ground for the step to be discussed in the next chapter: to discover *why* the differences occur at the points revealed by this refinement operation.

8

Experimental Evidence

THE PROBLEM

It would not seem too difficult to find out whether a new medical treatment, or a new method of teaching, or a legal innovation accomplishes its aim. All one would have to do is to compare persons who received the treatment with persons who did not receive it. Consider, however, the following comparisons of the driving record of youngsters who had taken a driving course in high school before obtaining their license with the driving record of youngsters who had not taken such a course. Of those who took the course, we will assume that 5.2 percent had an accident during their first year of driving; those who did not take it showed a rate of 8.8 percent (Figure 8-1).

It is the type of statistics we are likely to find in a booklet that recommends such driving courses. We are supposed to and probably do read it to mean that because youngsters took such a

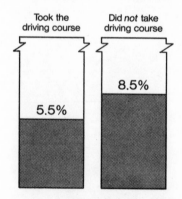

FIGURE 8-1
*Proportion of Youngsters with Accidents
During Their First Year of Driving*

course they are half as likely to have an accident when they begin
to drive than are the youngsters who did not take it. What we
may not be told, perhaps because the author does not consider it
important, is that the decision about whether or not to take the
course was left to each youngster. Once we are aware of this fact,
we will begin to suspect the result. It is possible and indeed not
unlikely that the nice, naturally careful youngsters are more likely
to have taken the course than are the daredevils. Even without
the course, the first group could have had fewer accidents. The
analysis in Figure 8-2 would bear these thoughts out.

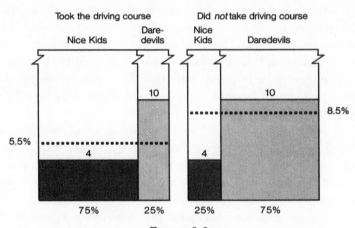

FIGURE 8-2
Course Effect on Daredevils and Nice Kids

Nice kids, this analysis would show, have fewer accidents (4
percent) than do daredevils (10 percent), even if they do not take
the course; daredevils have more accidents even if they take it.
Figure 8-1 is misleading because the youngsters who took the
course were different to begin with from the youngsters who did
not take it. More of the naturally careful drivers took the course,
more of the daredevils did not take it, and daredevils have more
accidents than do careful drivers. Because they were careful they

took the course, but—in this example at least—the course accomplished nothing.

There is only one safe way of avoiding such spurious comparisons: by using the so-called controlled randomized experiment.

THE CONTROLLED RANDOMIZED EXPERIMENT

Originally developed in the field of agronomics, we know the controlled experiment is best known from its frequent use in the widely reported testing procedures of new drugs or other new medical treatments. Patients who suffer from a particular ailment are invited to cooperate in the experiment; those who do are then separated randomly—that is by some sort of lottery—into two groups; one will obtain the new medical treatment and the other, the old traditional therapy. If the new treatment cures the disease more frequently, or faster, or safer, it will emerge as the superior treatment. Figure 8-3 presents the basic paradigm of the controlled experiment.[1]

The experimental method has the advantage of allowing for the application of what has been called the artichoke principle,[2] that is, it can proceed step by step in disentangling a complex causal problem. Two of the simpler devices for the stepwise treatment of causal problems are: (1) varying one or a few of the controlled causal factors at a time, keeping the other ones constant; (2) arranging for additive and linear effects of the controlled variables.

---

[1] We will skip here the technical niceties of experimental design. They can be found in any number of texts—for example, D. R. Cox, *Planning of Experiments* (New York: Wiley, 1958); also K. A. Brownlee, "The Principles of Experimental Design," *Industrial Quality Control*, vol. 13, 1957, pp. 1–9. See also S. A. Stouffer, "Some Observations on Study Design," *American Journal of Sociology*, vol. 55, 1949–1950, pp. 355–361.

[2] Herman Wold, "Causal Inference from Observational Data," *Journal of the Royal Statistical Society*, vol. 119, 1956, pp. 28–60.

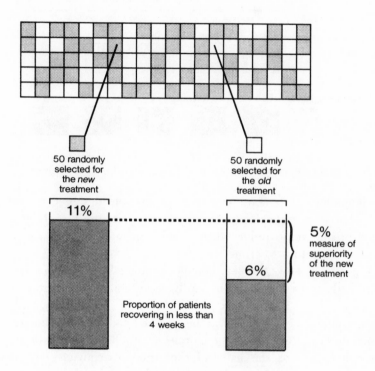

<figure>
FIGURE 8-3
The Randomized Controlled Experiment
(Recovery Rates after Two Different Treatments)
</figure>

THE MIRACLE OF RANDOM SELECTION

Selecting units at random, that is, by some lottery, from a group or a list means that each member has the same chance of being selected. This has the effect of making the two groups—the selected persons and the remaining ones—practically indistinguishable, except for usually slight variations the magnitude of which can be estimated.

In Figure 8-4, for instance, are five groups of ex-convicts who

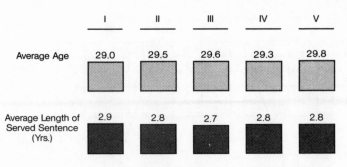

FIGURE 8-4

Effect of Random Separation of Ex-Convicts into 5 Groups

participated in a famous large-scale experiment designed to find out whether giving ex-convicts financial support after their release from prison would reduce their recidivism rate or at least slow down their return to crime.[3]

Figure 8-4 shows the near equality produced by random selection with respect to certain easily ascertainable demographic characteristics. That near equality, however, extends to *every* conceivable characteristic, whether defined and recorded beforehand, or only after the selection, or never. If it had been important enough to record the hair color of these ex-convicts, or the proportion among them who had ever been operated upon, or the proportion suffering from nightmares, the respective percentages would have been equally similar.

We are not unfamiliar with this effect of random selection. We know, for instance, that a true roulette wheel tends to distribute its rollings of the ball more or less equally between red and black numbers, even and odd ones, etc. The election polls produce perhaps the most astounding effect of random selection. By interviewing a random sample of some 1,500 prospective voters,

[3] Peter H. Rossi, Richard A. Berk, Kenneth H. Lenihan, *Money, Work, and Crime* (New York: Academic Press 1980), p. 77. On the dangers of misinterpreting even such a near perfect experiment, see H. Zeisel, "Disagreement Over the Evaluation of a Controlled Experiment," *American Journal of Sociology*, vol. 88, 1982, pp. 378–394.

we can learn with considerable precision the voting intentions of the remaining 60-odd million voters who were not interviewed, because random sampling ensures that the two groups are alike. To be able to do this, albeit within the now familiar error range of ±3 percentage points or so, strikes me ever again as miraculous.

To return to randomization in experiments. Making experimental and control group practically indistinguishable at the outset of the experiment has this invaluable advantage: if the treated group subsequently shows an effect that is not found in the control group, we are entitled to conclude that it was the experimental treatment that produced the effect. It *must* have been the treatment, because it constituted the only difference between the two groups which, prior to the experiment, had been alike in all other respects.

For this reason controlled experiments are regularly used in agriculture and industry, as well as in medicine, first with animals and eventually with humans. Controlled experiments involving humans are extremely rare for a variety of reasons, one of which is that human beings seldom hold still for experiments. The major obstacle, however, is that experimentation, by definition, involves discrimination; and if the issue is serious, discrimination poses problems of propriety.

DISCRIMINATION

Many considerations tend to reduce what at first may look to be an unsurmountable obstacle. There are times when the treatment is in short supply: this is occasionally the case with respect to pharmaceutical drugs in the initial stage of their development, which makes it impossible to supply the drug to all who need it. The Salk polio vaccine is an example. Some treatments are permanently in short supply, as, for instance, halfway houses in which ex-convicts learn to adapt to the nonprison world, or the once extravagant medical treatments of kidney dialysis. To select in such situations the recipients of the treatment by lot might be the fairest way of allocating short supply even if it is not part of an experimental design.

If the experimental treatment involves discomfort or even danger, as, for instance, in the testing of the healing power of two competing surgical treatments, still another problem arises. The participants must be advised and must consent. Again, however, the problem is less formidable than it may first appear. An experiment is needed only if it is uncertain which of the competing treatments is superior, thereby leaving it unclear whether it is preferable to be in the experimental group instead of in the control group. And whatever discrimination may be revealed through the experiment will be removed once the experiment is ended; the treatment that emerges as superior will become the general treatment, thus not only ending the temporary discrimination but improving subsequent treatment for all.

Medical experimentation, of course, poses more serious problems than do, for instance, experiments in teaching methods. Some of the first controlled experiments in that latter field were conducted without much ado in the army, where one company could be ordered to watch an indoctrination film while another was used as a control group.[4]

GENERALIZING FROM AN EXPERIMENT

The basically simple structure of the controlled experiment may convey the impression that such properly conducted experiments yield a vast fund of knowledge. They may, but often the yield is modest. Every experiment is limited in that it is conducted with certain subjects, in a certain place, at a certain time. The question of how far one may generalize from such a unique experience is forever puzzling. That one indoctrination film produces a measurable effect may allow little generalization; how is one to know

[4] Carl I. Hovland, Arthur A. Lumsdaine, and Fred D. Sheffield, *Experiments on Mass Communication*, Vol. III of *Studies in Social Psychology in World War II* (Princeton, N.J.: Princeton University Press, 1949).

whether the next film will be equally effective? A controlled experiment with two different instructions from the judge to the jury on the law of insanity in a case of housebreaking produced clearly different jury verdicts.[5] Does this mean that the difference will persist if the charge is murder? The problem of generalizing exists also in the natural sciences, but since these sciences have a more coherent structure, their experiments allow wider generalizations. The social sciences for the most part as yet lack such coherence and therefore offer fewer opportunities for broad generalization.

The question about the reach of an experiment can never be answered by the experiment itself; it cannot give assurance that its result will remain the same if conducted at another place, or under different circumstances. Such assurance of relative independence from place and time can come only from outside knowledge. If a drug is found to be effective with patients in Colorado in 1983, knowing about the relative independence of drug effectiveness from local conditions will encourage us to infer that the drug will be effective also elsewhere and at any later time. The situation will be different if we know something about the dependence of the experimental treatment upon local conditions. The time elapsed before an egg becomes hard-boiled is known to vary with the altitude at which the water boils. The known altitude effect can be factored into the experimental formula and thus increase the range of its validity.

In any event, extrapolation of an experimental result is an operation that must be reasoned and by definition entails risks, because in the new setting unknown forces could come into play. The best way to reduce the risk of overbroad generalization is to build many different conditions into the original experimental design or, if necessary, repeat the experiment under varying conditions.

[5] See Rita James Simon, *The Jury and the Defense of Insanity* (Boston: Little, Brown, 1967).

DESIGNATING EXPERIMENTAL AND CONTROL SUBJECTS

How, as a practical matter, does one perform random separation?
One begins with a list of the persons or units that will participate
in the experiment. The list—the *sampling frame*—might already
exist, or it might develop during the course of the experiment, as
in the experiment involving prisoners who will be released during
the experimental period. Next, a lottery device must be used to
ensure that the separation is left to chance. A page or two of
random numbers from any elementary statistics text will always
provide the perfect tool. Often, however, especially when the
experimental group constitutes but a small fraction of the total
list, what is called a systematic sample will suffice; selecting
randomly the starting unit and subsequently every nth unit from
the list usually ensures random selection. If the list develops
during the experiment, as in the convict example, extraordinary
precautions must be taken, lest the random assignment is subverted
by assistants of the experimenter, who may be tempted to assign
a beneficial treatment to a favored person.

If there is no list, one must make one. In a fertilizer experiment,
for instance, a large field must be divided into a "list" of small
plots which then, with the help of random numbers, are designated
as experimental and control plots.

As any sampling operation, the efficacy of separating experimental from control subjects can be enhanced by what is called
stratification. Instead of applying the random selection process to
the total list, one can break it down into more homogeneous parts
which potentially may react differently to the experimental treatment. If, for instance, male and female participants had been
involved in the ex-convict experiment, one might make a separate
list for each. Such stratified random selection ensures that the
proportions—in this case men and women—will be precisely the
same in the experimental and the control group, which procedure

may (but need not) increase the statistical power of the experiment if the sex of the participant indeed makes a difference. If it does not, stratification will have been irrelevant, but never harmful.

STATISTICAL ERROR

Although random selection generally results in providing comparable groups, it does not always do that; in some cases it may not. The trouble is that one never knows whether or not this has happened in a particular case. It may happen that a sample experiment may fail to show an effect which in reality exists and would have appeared had the whole list participated in the experiment. Statisticians call it the type II error. Conversely, an experiment may show an effect which actually does not exist, the type I error.

One can never be certain whether a particular random separation misbehaved, but statistical theory allows us to compute with some exactness the likelihood of such an event. In the main, that likelihood is a function of the size of the sample; the greater the sample, the smaller the likelihood that it will mislead.

THE NATURAL EXPERIMENT

Occasionally the designer of a controlled randomized experiment finds his task half done when he discovers that administrative routine or other more or less natural causes have provided him with the needed random assignment to different treatments.

The first study ever made of how criminal sentences varied from judge to judge profited from such an arrangement. In 1933, Gaudet found that criminal cases were randomly assigned to the several judges of the New York Municipal Court. He was thereby assured that whatever differences in sentencing patterns he found among these judges, he could attribute to their different sentencing policies, since the cases before them were for all practical purposes identical.[6]

A somewhat less transparent but essentially equivalent situation presented itself for a study of the relation between the menstrual cycle and the frequency of driving accidents. When the frequency of accidents is plotted against the time of menstruation a surprisingly shaped curve emerges.

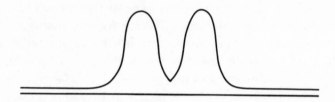

Upon investigation, this curve turned out to be the composite of two easily identified separate curves; one for parous women (those who had given birth) and one for nonparous women. The one group had the accident peak immediately after their period, the other immediately before it.[7]

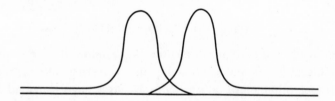

In this example, the prior randomization process is only inferred,

[6] F. J. Gaudet, G. S. Harris and C. W. St. John, "Individual Differences in the Sentencing Tendencies of Judges," *Journal of Criminal Law and Criminology*, vol. 23 (1933), p. 811. Also H. Zeisel and T. Callahan, "Split Trials and Time-Saving: A Statistical Analysis," *Harvard Law Review*, vol. 76, 1963, p. 1606ff.

[7] From Katherina Dalton, *Menstruation and Accidents* in W. Haddon, E. Suchman, D. Klein (eds.), *Accident Research*. (New York: Harper & Row, 1964).

not explicit; hence it is less certain. The inference is that women do not drive more frequently or along more dangerous routes during their menstrual cycle.

In searching for answers to the question of how much human character is determined by inherited genes (nature) and how much by life experience (nurture), scientists have found identical twins, who have identical genetics, to offer a convenient study situation, even though the differential "treatment" here is designed by life, not by the experimenter.

THE HALF-A-LOAF EXPERIMENT

Every so often the ideal controlled randomized experiment cannot be carried out because factual or legal obstacles stand in the way.

If, for instance, one wanted to find out what difference it makes whether a case is tried before a jury or before a judge without a jury, one might toy with the obviously impossible idea of trying each case twice, once with jury and once without. If one wanted to find out whether smoking increased mortality from cancer, one might think of the impossible idea of providing a randomly selected group of youngsters with an unlimited supply of free cigarettes for the rest of their lives and forbidding smoking to a comparable control group.

In such seemingly hopeless situations, a modification of the original design will at times make the experiment possible, albeit, as a rule, at the price of weakening the power of the experiment.

The jury trial dilemma, for instance, was resolved by asking the judges who preside over jury trials to tell us how they would have decided these cases had they tried them without a jury. Since both jury and judge hear the same case, the experiment *is* controlled. The design has only one, as it turns out negligible, weakness: the judge's "verdict" is not real.[8]

The murderous slant of the ideal smoking experiment can be

[8] See H. Kalven, Jr., and H. Zeisel, *The American Jury*, 2d ed. (Chicago: The University of Chicago Press, 1971).

removed by a different route, namely, by reversing the experimental treatment. Instead of encouraging a random half of youngsters to smoke, the experiment would encourage randomly selected youngsters not to smoke. To the extent to which that effort is successful, the respective morbidity and mortality rates would indicate the effect of abstaining from smoking.

At times the ideal experiment is bound to fail because the experimental situation would bias the result. In such a situation a shift of the "treatment" to the control group may resolve the difficulty. If one wanted to test the effectiveness of a television commercial, asking viewers to watch the program that carries the commercial might make them self-conscious and thus biased in their reaction. To avoid bias, it was suggested that the control group, under some pretext, be asked to watch another television program aired at the same time, with the sole purpose of preventing the group from watching the experimental commercial.[9]

In civil litigation there is something called pretrial conference, in which the judge prior to trial confers with opposing counsel and their clients to learn whether the case can be settled. Although a great many cases have been settled at these conferences, there has been some doubt about whether they would not have been settled anyway, with or without pretrial conference.[10]

To test this hypothesis we proposed to the New Jersey courts a controlled experiment: pretry a random half of the filed civil cases, withhold pretrial from the other half, and then compare the settlement rates of the two groups. The court agreed to the experiment but with a modification. To deprive litigants of their right to pretrial seemed to violate constitutional rights. The court suggested a change in the experimental design: instead of depriving

[9] Irving Towers, Leo Goodman, and Hans Zeisel, "A Method of Measuring the Effects of Television Through Controlled Field Experiments," *Studies in Public Communication*, University of Chicago, no. 4, 1962, pp. 87–110.

[10] H. Zeisel, H. Kalven, Jr., and B. Buchholz, *Daley in the Court* (Boston: Little, Brown, 1959), p. 143. The experiment was conducted and analyzed by Maurice Rosenberg, *The Pretrial Conference and Effective Justice* (New York: Columbia University Press, 1964).

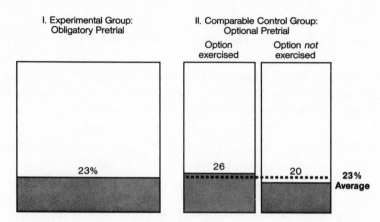

FIGURE 8-5
*Proportion of Civil Law Suits Settled before Trial with and without
Pretrial Conference*

the litigants in the control group of their right to pretrial, they
should have one if they demanded it. Making the treatment
optional, however, involved the danger that all or most of the
litigants in the control group would make use of their option, in
which case there would have been no experiment, since the
litigants in both groups would have their cases pretried. As it
turned out only one half of the control group wanted pretrial,
with the results shown in Figure 8-5.

One must resist, incidentally, the temptation to compare the
experimental group with the part of the control group that had
no pretrial—the last bar in Figure 8-5. The original random
division separated I from II, and it is therefore necessary to
compare the settlement ratios for I with that of the average for
the entire group II. Once random selection has assigned a subject
or unit to a group, it must stay there forever, whether or not it
receives the treatment.

The price to be paid for changing the ideal but impossible
experimental design into a feasible one may take various forms.
There is the risk that the required voluntary cooperation may

vitiate the experiment. The main price, however, is the weakening of the reformulated hypothesis that is being tested. In the pretrial experiment, the optional pretrial rather than the elimination of pretrial was tested. In the smoking experiment, giving up smoking would be tested rather than smoking. Often the reformulation reduces the power of the experiment because it replaces some design elements through simulated ones, as was done in the jury study.

SUMMARY

The controlled randomized experiment offers singular advantages. By controlling the treatment, and by random separation of experimental and control subjects, relatively precise inferences can be drawn about the effectiveness or ineffectiveness of the treatment. Controlled experiments involving humans, however, pose problems of logistics as well as propriety, and therefore require careful design and management. Such experiments, moreover, tend to have a relatively narrow focus. Nevertheless, if certain issues require correct judgment, the controlled randomized experiment is an indispensable research device. Often, therefore, it will be worthwhile to modify an experimental design if that helps to retain the essential requirement of prior randomization. The precision of the controlled experiment must not be allowed to obscure a major problem of interpretation: how far in space and time can its results be generalized? That decision cannot be derived from the experiment itself, only from outside knowledge about the tested connection.

9

Analysis of Nonexperimental Data

A DIFFERENT PROBLEM

If a treatment is not administered in controlled, experimental fashion, the task of analysis changes radically. It is no longer permissible to assume that experimental and control groups were indistinguishable prior to the treatment. On the contrary, one must start with the assumption that the observed effect difference between the treated and the not-treated group—or the lack thereof—may be the compound effect of the treatment *and* of any number of differences that distinguished experimental and control groups *prior* to the treatment.

On the surface, the survey cross tabulation is indistinguishable from the tabulation of an experiment. It too compares those exposed to the potential cause with those who have not been exposed, but the comparison is made retrospectively, that is, without prior randomization of the two groups—and this makes a big difference. Without prior randomization there is no way of categorically stating that prior to the experimental exposure, the two groups had been interchangeable—an indispensable requirement for ascribing the resulting effect with certainty to the exposure. The survey cross tabulation is therefore in need of reassurance that in spite of the lack of prior randomization, the two groups had been originally interchangeable.

There is an old Chinese statistical joke to the effect that the people who are visited by a doctor have a considerably higher mortality rate than those who have been spared such a visit. Extreme examples have the advantage of clarifying both the problem and its solution. Obviously, what is needed here is to make the visit–no visit comparison separately for those who had been sick and those who had not been sick prior to the event; we would then quickly find out that the latter group could safely be

eliminated altogether, because it had no professional visits from
doctors. The general prescription here is to split the cross tabulation into segments and make the exposure–nonexposure comparison separately for each of the subgroups of the total population.
And the question now is, along what lines should these subgroups
be formed? Again, the doctor example shows the way: We must
group the people in such a way that we remove whatever impeded
their interchangeability prior to the exposure. Our difficulty was
that practically all people in the group who had a doctor visit
must have been sick, while those who had no doctor visit probably
included some sick persons but also all those who had no need
for a doctor. By making the comparison separately for the sick
and the healthy, we remove the biasing inequality in the original
group. Incidentally, not every inequality will bias the result. If
one group in the doctor example had contained more redheads
than did the other, such inequality would in all probability not
have mattered. But if one group had been older than the other,
it would have, because older people will, on the whole, die sooner
than younger ones.

The trouble with this procedure of fractioned analysis is that
however far it is carried, it never guarantees cure; there is no
way of knowing for certain that all the hidden factors that will
render a causal inference spurious have been eliminated, however
plausible the result. Thus statistical confidence tests may be
applied to survey data only with discretion.[1] Ultimately, what
Michael Polanyi has called personal knowledge must here play an
important role.[2] But it would be a mistake to conclude from these
analytical difficulties that the survey is generally inferior to the
experiment. For one thing, survey data are gathered in their
natural settings where they have developed normally without

[1] See Hans Zeisel, "The Significance of Insignificance Differences," *Public
Opinion Quarterly*, vol. 19, 1955, p. 319.
[2] *Personal Knowledge* (London: Routledge, 1958).

interference on the part of the manipulating scientist.[3]

The fact that paroled offenders show a lower recidivism rate than do offenders who served their full prison term does *not* allow us to attribute this difference to the parole treatment; these prisoners may have been paroled *because* they were better risks to begin with. Data such as these that emerge from the ordinary keeping of administrative or business records are called *observational* data, to distinguish them from *experimental* data obtained through randomized controlled experimentation.

At times a new treatment is introduced "experimentally," i.e., for the purpose of trying it out, that does not have the safeguards of randomized control. The data from such "quasi-experiments" as they are called, although more tractable, share the deficiency of observational data which force us to sort out the effects of the treatment itself from those attributable to the original differences between the treated and the not-treated group.

The rest of this chapter pursues the logic and practice of this sorting-out process in some detail.

FULL EXPLANATION

We begin by analyzing the data we looked at before in Table 7-4, so unflattering to the male ego, which shows that female drivers had fewer accidents than do male drivers. While it is possible that women are simply better or at least more careful drivers than men, the analytic task is to see whether their lower accident rate may have another explanation. Women, for instance, could have fewer accidents than men do because they drive less.

Table 9-1 presents schematically the first step in this analysis by exploring whether women do indeed drive less than men. Male and female drivers were divided into two subgroups according to

[3] In the study of animal behavior a similar dichotomy exists: The laboratory experiment is but one of its research tools. Systematic observation of animals in their natural settings, something aided by the observer's experimental design, a method developed primarily by Konrad Lorenz and Nikolaas Tinbergen, has emerged as a most powerful research procedure.

TABLE 9-1
Amount of Driving, by Sex of Driver

	Men %	Women %
Drive annually:		
More than 10,000 miles	71	28
10,000 miles or less	29	72
Total	100	100
(Number of persons)	(7,080)	(6,950)

whether they annually drove more than 10,000 miles or 10,000 miles or less. Men, it turns out, do drive more than women. Of the male drivers 71 percent are in the more-than-10,000-miles group as compared with only 28 percent of the female drivers.

Table 9-2 represents the next analytic step: to what extent does

TABLE 9-2
Automobile Accidents of Male and Female Drivers,
by Amount of Driving

	Male Drivers		Female Drivers	
	Drove More than 10,000 Miles	Drove 10,000 Miles or Less	Drove More than 10,000 Miles	Drove 10,000 Miles or Less
Percent of persons in that group with at least one accident	52	25	52	25
(Number of persons = 100%)	(5,010)	(2,070)	(1,915)	(5,035)
	(7,080)		(6,950)	

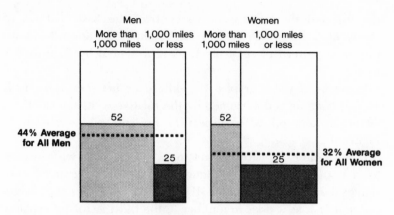

FIGURE 9-1
Automobile Accidents of Male and Female Drivers,
by Amount of Driving

the fact that men drive more than women account for the difference in their accident rate?

If one compares men and women who drive within the same annual mileage range, the difference in their accident rates disappears; it is 25 percent for both if they drive 10,000 miles or less, and 52 percent for both if they drive over 10,000 miles. In our simplified example, driving more miles fully explains why men have more accidents than do women: not because they are poorer drivers, but because they drive more.

If the injection of the mileage factor had failed to explain the differential accident rates between men and women, one might have tested other competing hypotheses, e.g., that men have more accidents because they are more likely to drive during rush hours in fast and congested traffic. To the extent that the competing hypotheses fail to provide an explanation for the accident differential, the sexist explanation, that women drive better, will gain ground.

Figure 9-1 presents the two preceding tables in graphic form.

It will help to clarify the nature of this explanatory function of the injected third factor if we contrast the basic pattern of *explanation,* for which we have an example in Figure 9-1, with the basic pattern of *refinement,* as represented by Figures 8-1 to 8-3.

In terms of these graphs, the difference between *refinement* and *explanation* is determined by the relative widths of the third factor (c_1 and c_2) with respect to (a_1 and a_2), as shown in Figure 9-2.

If the ratio c_1/c_2 with respect to (a_1) is the same with respect to (a_2), as in number 2 of Figure 9-2, the third factor (c) only refines the original correlation. If the ratio c_1/c_2 differs as between (a_1) and (a_2), as it does in number 3, the third factor (c) explains the correlation, provided (c) is also related to (b), the variable that is to be explained.

PARTIAL EXPLANATION

It will be the exception rather than the rule that the third factor provides a complete explanation. In the real world, the third factor is more likely to provide only part of the explanation. In a survey on factory absenteeism, for instance, it was found that married women had a higher rate of absenteeism than did single women, as shown in Table 9-3.

The investigator was interested in finding out what was involved in being married that caused the difference—having another bread winner in the family; being on the average older and more tired than the single women; or being confronted with the demands of more housework? Table 9-4 tested this last hypothesis.

Married women, it turned out, have more housework. Table 9-5 now presents the absenteeism rates for the four groups of housewives of Table 9-4.

Table 9-5 suggests that the major part of the increase in absenteeism of married women is indeed caused by the demands of housework, not simply by marriage. The absenteeism rate among married women is almost as small as that of single women

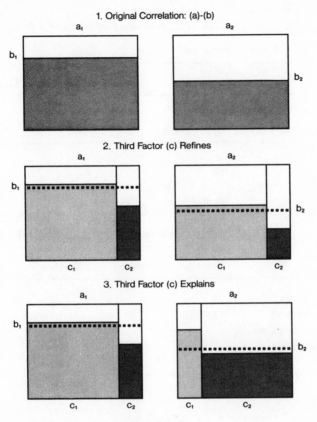

FIGURE 9-2
The Third Factor Refines or Explains the Correlation

if both have little or no housework; and absenteeism among single women is almost as high as that of married women if they too have a great deal of housework.

In a way, of course, it *is* marriage that increases absenteeism, because, as Table 9-4 shows, marriage brings with it more housework. But the explanation is not complete. Among the women having much housework, as well as among the women having little or no housework, the married women show a higher rate of

TABLE 9-3

Absenteeism Rates of Women Workers by Marital Status

(Percent of working days absent)

	Married	Single
	6.4	2.3
(Number of working days = 100%)	(6,496)	(10,230)

TABLE 9-4

Marital Status and Housework

	Married Women		Single Women	
	Number	Percent	Number	Percent
Have much housework	(5,680)	88	(1,104)	10
Have little or no housework	(816)	12	(9,126)	90
Total	(6,496)	100	(10,230)	100

TABLE 9-5

Absenteeism Rates of Women Workers, by Marital Status and Amount of Housework

	Percent of working days absent	
	Married Women	Single Women
Have a great deal of housework	7.0	5.7
Have little or no housework	2.2	1.9
Total average*	6.4	2.3

* See line 1 in Table 9-3.

absenteeism, the remaining difference being a fraction of the original difference shown in Table 9-3. This unexplained residual must have reasons other than housework, but related to marriage; students may come up with some good guesses.

SPURIOUS CORRELATIONS

We turn now to a series of examples of a slightly different sort. Again the third factor explains the original correlation, but it explains it by invalidating it at the same time.

Again we begin with a correlation in which being married is associated with a difference in behavior, in this case with the amount of candy eating. Fewer married women eat candy than do single women. Table 9-6 was obtained from a survey of candy eating in which 3,009 persons were interviewed. Again, marital status is the correlated mystery attribute.

TABLE 9-6
Women Eating Candy, by Marital Status

	Single	Married
Eat candy	75%	63%
(Number of cases)	(999)	(2,010)

The same sample analyzed by age instead of marital status yielded Table 9-7.

The first table shows that fewer married women eat candy; the second table shows that fewer older women eat candy. Since it is

TABLE 9-7
Women Eating Candy, by Age

	Up to 25 Years	25 Years and Over
Eat candy	80%	58%
(Number of cases)	(1,302)	(1,707)

TABLE 9-8

Women Eating Candy, by Age and Marital Status

	Up to 25 Years		25 Years and Over	
	Single	Married	Single	Married
Eat candy	79%	81%	60%	58%
(Number of cases)	(799)	(503)	(200)	(1,507)

a commonplace that married people are on the average also older, however, the question arises: do fewer married than single people eat candy because they are *married*, that is, because the husband got his girl and no longer needs to bring her candy; or do married people eat candy less frequently because they are on the average *older*, have outgrown their sweet tooth, and are more afraid of gaining weight? We might guess either way, but the truth cannot be learned from Tables 9-6 and 9-7. The data must be broken down, as in Table 9-8, which allows us to see simultaneously the relation between candy eating, age, and marital status.

Figure 9-3 shows this relation graphically.

Table 9-8 and Figure 9-3 reveal that the relation between being married and eating less candy is fully explained by the fact that married people are, on the average, older than single people, and because older people eat candy less frequently. If married and single people of equal age are compared, the association between marital status and candy eating disappears; the figures clear the husband of all suspicion—or, at least, of this one.

PARTLY SPURIOUS CORRELATION

In the candy-eating example, it was not difficult to discover the true correlation, because everybody knows that married people tend to be older than single people. The choice of the potential third explanatory variable is not always so obvious.

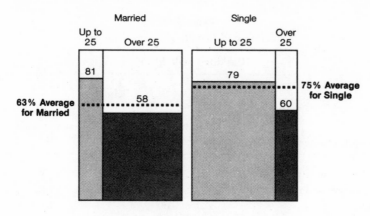

FIGURE 9-3
Candy Eating, by Age and Marital Status

An inquiry into the characteristics of unemployed workers during the Great Depression of the 1930s revealed a strong negative correlation between the amount of schooling and the duration of unemployment.[4]

Table 9-9 refers to the subgroup of unskilled male black workers for whom the association was particularly strong.

Workers with more schooling seem to have a considerably better chance of being unemployed for only a shorter period of time, which suggests that more schooling provides protection against long unemployment. But when age is introduced into the analysis (Table 9-10), an unexpected new relation emerges.

Table 9-10 reveals that the workers with less schooling are on the whole also older. The correlation between schooling and unemployment almost disappears, and age emerges as the main factor that determines the length of unemployment, regardless of the amount of schooling. The spurious correlation of schooling and unemployment appeared only because the better-educated workers were, on the average, of a younger generation, a phe-

[4] Katherine D. Wood, *Urban Workers on Relief*, Part I, Research Monograph IV (Washington, D.C.: U.S. Government Printing Office, 1936).

TABLE 9-9

Schooling and Length of Unemployment

(Unskilled male blacks)

	Schooling	
	Up to 5 Years	5 Years and More
Length of unemployment	%	%
Less than 2 years	47	52
2 years and more	53	48
Total	100	100
(Number of cases)	(6,054)	(6,039)

nomenon characteristic for all countries in which the average level of education is on the increase.

Again, age does not provide a complete explanation. There remains a residual advantage accruing to those who have more schooling, which might properly be attributed to the benefits of education.

TABLE 9-10

Length of Unemployment, by Age and Schooling

(Unskilled male blacks)

	SCHOOLING			
	Up to 5 Years		5 Years and More	
	Up to 35 Years of Age	35 Years or More	Up to 35 Years of Age	35 Years or More
Length of Unemployment	%	%	%	%
Less than 2 years	58	42	60	44
2 years and more	42	58	40	56
Total	100	100	100	100
(Number of cases)	(1,823)	(4,231)	(3,241)	(2,798)

THE CORRELATION IS REVERSED

Table 9-11 presents findings from a study of credit risks in installment buying.

TABLE 9-11
Credit Risks by Price of Purchased Item *

	Price under $60 %	Price $60 and Over %
Bad credit risk	13	10
Good credit risk	87	90
Total	100	100
(Number of cases)	(4,303)	(4,088)

* Adapted from David Durand, *Risk Elements in Consumer Installment Financing* (New York: National Bureau of Economic Research, 1941.)

Purchasers of high-priced items appear to be better credit risks than are those who buy lower-priced items. When the size of the down payment is added as a third factor, the more complicated picture of Table 9-12 emerges.

The correlation is now more pronounced. But its main determinant is the size of the down payment. A large down payment

TABLE 9-12
Bad Credit Risks by Price of Item and Size of Down Payment

	LOW-PRICED ITEMS		HIGH-PRICED ITEMS	
	Down Payment		Down Payment	
	Small	Large	Small	Large
Percentage of bad credit risks	14	6	15	8
(Number of cases)	(3,655)	(648)	(890)	(3,198)

reduces the credit risk. Still, in each down-payment category, the purchasers of lower priced items are now the better credit risks than are the purchasers of high-priced items (14 vs. 15 and 6 vs. 8).

SPURIOUS NONCORRELATION

In a survey of milk consumption there seemed to be no correlation between the income of a family and the amount of its milk consumption (Table 9-13).

TABLE 9-13
Milk Consumption by Family Income

	Above-Average Income	Below-Average Income
Weekly consumption, quarts	10.8	11.0
(Number of families)	(503)	(498)

This somewhat surprising noncorrelation was explained when family size was considered as a third factor (Table 9-14).

Milk consumption, as expected, is affected by family size; the

TABLE 9-14
Milk Consumption by Family Income and Size

	Below-Average Income		Above-Average Income	
	Three or Fewer in Family	Four or More	Three or Fewer in Family	Four or More
Weekly milk consumption, quarts	6.2	14.4	8.0	17.1
(Number of families)	(281)	(222)	(334)	(164)

larger family consumes more than twice as much milk as does the smaller family. In addition, however, families comparable in size, if they are in the higher income bracket, consume about one-third more milk than does their counterpart with below-average income: 8.0 against 6.2 quarts, and 17.1 against 14.4 quarts. Thus, after proper analysis, the original noncorrelation proves to be spurious; the correlation between family size and income obscures the correlation between income and milk consumption.

TRUE AND SPURIOUS CORRELATIONS

We have called some of the presented explanations "spurious," so it is time to define the term with some precision. To this purpose we will look at the two examples in which being married turned out to be correlated with a certain behavior, less candy eating in one example and more absenteeism in another.

1. Why do married women have a higher rate of absenteeism than do single women? Because married women have more housework and more housework results in greater absenteeism.
2. Why do married people eat less candy than do single people? Because married people are on the average older, and older people eat less candy.

In both examples the introduction of a third factor explained the original correlation. Yet, in the one case we accepted the explanation as true; in the other case we decided that the correlation with being married was *spurious*, suggesting that marriage had nothing to do with causing the differential behavior.

The difference comes from the different role played by the third, the explaining factor: more housework in example 1; and old age in example 2. In case 1, more housework is the result of being married and is, in turn, the cause of higher absenteeism. In symbols—the arrows point in each case from the cause to the effect—the relation would read as follows:

getting married → more housework → more absenteeism

The important point is that the relation between more housework and getting married cannot be reversed. To have more housework will *not* increase the likelihood of getting married.

In example 2, the position of the explaining factor, getting older, is different. Getting older *will* increase the likelihood of one's being married, and reduce the likelihood of one's eating candy. In symbols:

getting married ← getting older → eating less candy

Note the reversed position of the first arrow: Getting older is not only the cause of eating less candy but also the cause—not the effect of—getting married.

A correlation is true when the explanatory third factor is asymmetrically connected with the two factors of the original correlation; it is spurious when the connection with the two original variables is symmetrical. In case 1, the explanatory factor (in the middle) is the *result* of the one factor and the *cause* of the other; in case 2, the explanatory factor is the *cause* of both.

The final test of the merit of the distinction between true and spurious correlations is a pragmatic one: a factory manager, acquainted with the fact that married women stay home more often, might consider discouraging female employees from marrying. Would such a policy, assuming it succeeded, reduce absenteeism? The answer is yes. Remaining single would result on the average in less housework, and less housework means less absenteeism.

Suppose now, a candy manufacturer familiar with our survey results were to entertain a similar notion: that candy consumption would increase if girls were advised not to marry. Would this effort, again assuming it succeeded, lead to an increase in candy consumption? Here the answer is no. Failure to marry does not keep girls from getting older. And since only not getting older would keep candy consumption high, marrying or not marrying would make no difference.

Note that the distinction between true and spurious is not revealed by the *statistical* relations between the three variables. Whether the two arrows in the schematic presentation run parallel or point in opposite directions can be derived only from outside knowledge. We know from experience, not from the presented statistics, that getting married does not affect a woman's age, but it does affect the amount of her housework.

The theoretical and practical reasons for distinguishing between true and spurious correlations are now clear: only the true correlation reflects a causal connection; the spurious one does not.[5]

"BEFORE" AND "AFTER" COMPARISON

The same type of analysis is required if one wants to learn whether a certain institutional intervention had the expected effect. By comparing the situation before the intervention with the situation afterward, one hopes to catch the effect of the intervention. This design, known also as "interrupted time series," is the most frequently used quasi-experiment. If all other conditions, aside from the new treatment, remain unchanged during the before and after period, one might expect the observed shift in the time series to reflect the true impact of the new treatment. But whether these conditions remain unchanged or not can never be taken for granted and must be investigated in each case.

[5] The core of the preceding analysis was developed by Paul Lazarsfeld during his Vienna teaching days. A more elaborate treatment can be found in Patricia L. Kendall and Paul F. Lazarsfeld, "Problems of Survey Analysis," in Robert K. Merton and P. F. Lazarsfeld (eds.), *Continuities of Social Research* (New York: Free Press, 1950). In 1954 Herbert Simon took up the problem: *Spurious Correlation: A Causal Interpretation*, Cowles Commission Paper No. 89 (Chicago: The University of Chicago Press, 1954).

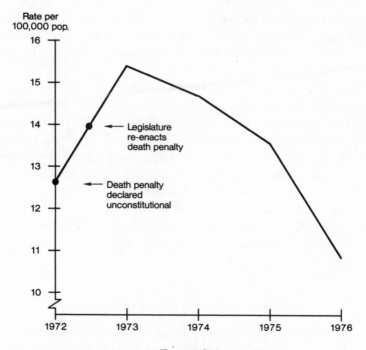

FIGURE 9-4
Homicide Rate, 1972–76, in Florida

In a court hearing in Florida some years ago, I presented evidence which showed that substituting the death penalty for life in prison as the sentence for murder would not increase deterrence and therefore not reduce the homicide rate. The state's attorney general in his cross examination presented me with the data charted in Figure 9-4, revealing a sharp decline of the Florida homicide rate following the reimposition of the death penalty.

The answer is given by Figure 9-5, which records the homicide rate in the states that did *not* reintroduce the death penalty.

These states show the same downward trend as does Florida, suggesting that the decline in homicides had more general causes and was unrelated to the reintroduction of the death penalty.

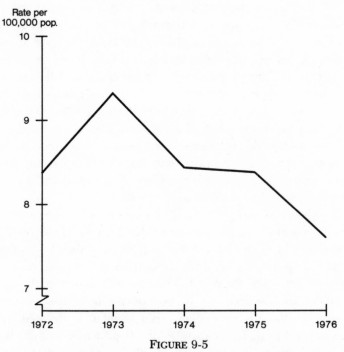

FIGURE 9-5

Homicide Rate in States Without Death Penalty, 1972–76

That insight could come only from the quasi-experimental comparison with a control group of states.

Interrupted time series raises the ever-lurking possibility that something else might have changed around the time of the interruption that affected the curve. Such unexpected interference need not necessarily produce a spurious effect, as in our example. The real effect could have been made invisible by a countervailing development.

How then is one to guard against faulty diagnosis suggested by the comparison of a time series before and after the intervening event? Figure 9-5 demonstrates one possible approach: look at comparable time series covering the same period in surroundings that are similar, except that there the interrupting event did not take place. The other approach begins by listing the factors known to influence the measured variable, the homicide rate in our example. Homicide is predominantly a male activity, of young men at that. If their proportion in the general population underwent some change at the time of the "interruption"—if, for instance, the soldiers came home from the wars—that fact alone would increase the number of homicides. In such a situation it would be preferable to measure the homicide rate per 100,000 men between 16 and 40 rather than per 100,000 persons of the general population. One must furthermore watch the rate of unemployment, of migration, and so forth.

If the observed change was radical enough and if one can think of few or no other factors that could have muddied the picture, the before and after test will be convincing. Infantile paralysis virtually disappeared after mass application of the vaccine. But it took some time until the reaction became sufficiently marked and permanent for the analysts to be convinced of the vaccine's success. Occasionally such time series proof is buttressed by a natural experiment. If the disease afflicts primarily persons who have not been vaccinated, the credibility of the time series evidence is enhanced.

LATERAL ANALYSIS

If we are concerned not with time series but with lateral, cross-section comparisons, an analagous analysis will often be helpful. Another effort to discover a deterrent effect of the death penalty concentrated not on the time dynamics but on the level of the homicide rate in states with and without the death penalty.

The homicide death rate in Michigan, Indiana, and Ohio is about the same. During a long period it was even identical, 3.5 per 100,000 population. Yet Michigan does not have the death penalty and its two neighboring states have it. The conclusion, however, that this equality suggests that the death penalty does not deter homicide is open to objections; there could be other differences between the three states that could hide the absence of an otherwise expected differential in the homicide rate. Table 9-15 investigates that possibility by examining a number of factors, which by their effect on the homicide rate could have obscured the death-penalty effect.

Michigan, the state without a death penalty, had no higher homicide rate than neighboring Indiana, even though it had a lower probability of apprehension and conviction, a higher unemployment rate, a larger proportion of blacks in the population, greater population density—all factors which should tend to increase the capital crime rate. On the other hand, it had a higher per capita police expenditure. Ohio had a lower homicide rate and a higher apprehension rate. On most of the remaining characteristics Ohio was in an intermediary position. This examination, therefore, suggests that the absence of an overall effect of the death penalty was not spurious.

TABLE 9-15

*Demographic Profile of Contiguous States**

(1960 Data)

	Michigan	Indiana	Ohio
Status of death penalty	NO	YES	YES
Homicide rate	4.3	4.3	3.2
Probability of apprehension	.75	.83	.85
Probability of conviction	.25	.55	.33
Labor force participation, percent	54.9	55.3	54.9
Unemployment rate, percent	6.9	4.2	5.5
Population aged 15–24, percent	12.9	13.4	12.9
Real per capita income, dollars	1,292	1,176	1,278
Nonwhite population, percent	10.4	6.2	9.8
Civilian population, thousands	7,811	4,653	9,690
Per capita government expenditures, dollars°	363	289	338
Per capita police expenditures, dollars°	11.3	7.6	9.0

* From D. Baldus & J. Cole, "A Comparison of the Work of Thorsten Sellin and Isaac Ehrlich on the Deterrent Effect of Capital Punishment," *Yale Law Journal*, vol. 851, 1976, pp. 170, 178.
° State and local.

SUMMARY

We often draw causal inferences from observational data, where the cause to be studied was *not* applied to comparable groups, originally separated through random selection. Such analysis is fraught with the danger that the differential effect, or its absence, was caused by factors that distinguished the groups prior to the impact of the studied cause.

Such analysis of observational data must make every effort to subdivide the groups in a fashion that will make equality between the groups more likely.

This analysis remains forever fragile, but it can gain strength if it becomes part of a broader analytical effort in which a variety of independent approaches all point toward the same conclusion.

10

Regression Analysis

Regression analysis is one of the powerful statistical tools that deal with the relationships among numerical variables representing such characteristics as age, height, income, and family size. This analysis allows us to estimate the average value of one variable (the "dependent" one) from the known value of other variables (the "independent" ones), which are correlated with it.

THE SCATTER DIAGRAM

The operation begins by asking to what extent, in the simplest form, two variables are associated or correlated. Plotting the data in the form of a so-called scatter diagram gives the first rough answer. Figure 10-1 is such a diagram which records for a sample of men the height of their fathers together with their own height.[1] Each dot represents one man. Its distance from the bottom line, measured along the vertical scale, indicates his own height. Its distance from the left-hand margin, measured along the horizontal scale at the bottom, indicates the height of his father. The diagram shows that generally the men who had taller fathers are themselves taller. We speak of a positive association or correlation. If one variable declines as the other grows, we speak of a negative association or correlation.

THE CORRELATION COEFFICIENT

The degree of association or correlation between two such variables is often measured by the so-called coefficient of correlation, or r for short. It is given a positive value (+) if the correlation is

[1] The data were collected and first published by Karl Pearson, a disciple of Sir Francis Galton (1822–1911) the British scientist who invented the basic methods discussed in this chapter.

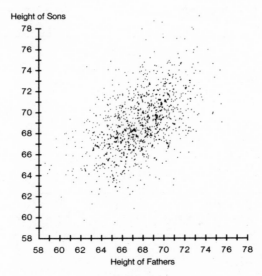

FIGURE 10-1
Scatter Diagram of Two Variables: Height of Sons vs.
Height of Their Fathers

positive and a negative value (−) if it is negative. *r* varies between
+1.0 and −1.0 (perfect positive and perfect negative correlation)
with zero at the midpoint indicating the absence of correlation.

Figure 10-2 shows by way of four examples how the shape of
the cloud of dots roughly indicates the magnitude of the coefficient
of correlation. It measures the degree of clustering around a line
that may or may not emerge with some clarity from the scatter
diagram. In panel one, comparing the I.Q.s of identical twins who
grew up together, such a line emerges with great clarity, the
clustering is modest; the correlation coefficient is correspondingly
high (+0.95). The second panel to the right reproduces our
Figure 10-1; a "line" is still visible but not too clearly; the
correlation coefficient is 0.5. For the remaining panels the coefficients are 0.3 and zero, the latter signifying the absence of
relationship.

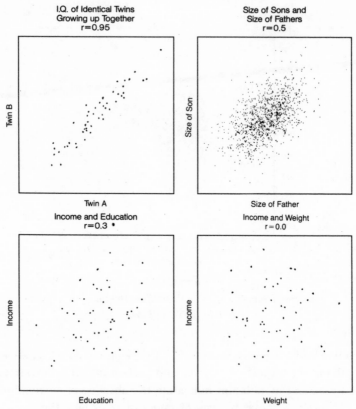

FIGURE 10-2

Four Scatter Diagrams with Their Respective Measures
of Correlation (r)

(Standardized for Mean and Standard Deviation)

We now turn to Figure 10-3, again the diagram of Figure 10-1, this time with some additional marks. They show how regression analysis allows us to estimate the average value of the dependent variable from our knowledge of the independent variable. Figure 10-3 is divided into vertical stripes, each representing to the nearest inch the height of the father. The points in that stripe represent the sons of the fathers of that particular

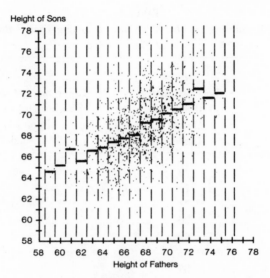

FIGURE 10-3
Estimating the Son's Height from the Height of the Father

height. And the horizontal little line in each stripe denotes the mean height of these sons. These measures form an almost straight, upward sloping line. The line, for reasons presently to be explained, is called the *regression line*. A father seeking the best estimate of the eventual height of his future son must enter Figure 10-3 at the bottom scale at his own height, move upward until the regression line is met, and then read the best estimate of his son's eventual height by moving horizontally to the scale at the left margin of the graph.

REGRESSION TOWARD THE MEAN

Table 10-1 contains this information, for each height, to the nearest inch, of the father.

Table 10-1 confirms, first, what we knew all along; there is a positive correlation between heights of the fathers and heights of

TABLE 10-1

Results of Estimating the Height of Sons, Knowing Their Father's Height and That r = 0.5

(a) Father's Height Inches	(b) Best Estimate of Son's Height Inches	(a) − (b) Difference between Height of Father and Best Estimate of Son's Height Inches
76	73	−3
74	72	−2
72	71	−1
70	70	−0
68	69	+1
66	68	+2
64	67	+3
62	66	+4
60	65	+5
58	64	+6

the sons. As the father's height increases from the bottom of column *a* to the top, so does the son's (*b*). Accompanying this increase is a remarkable pattern: for all fathers whose height is above the average, the best estimate of the son's height is relatively smaller, as shown by the upper half of the third column; for the fathers of below-average height, the son's height is relatively greater, as shown by the lower half of the third column. Thus for both the above- and below-average fathers, the corresponding heights of the sons is closer to the mean.

This movement of regression toward the mean, an inextricable part of the total phenomenon, prompted its discoverer, the British scientist Francis Galton, to name the line of best estimate the *regression* line. Correspondingly, the regression toward the mean is often called the *regression effect*. Simply stated, in any scatter diagram with positive correlation, subjects who are well above the average in the independent variable tend to be also above average in the dependent variable—but not by quite as much. And similarly, subjects well below the average in the independent variable tend still to be below the average in the dependent variable—but again not by as much.

THE REGRESSION FALLACY

Insensitivity to the regression effect may at times lead to false inferences. The danger lies in looking only at part of the picture and not at the whole. Suppose that we consider the correlation between the results from two consecutive tests of a group of students regarding their ability to perform in a particular skill. Assume that we obtain a correlation as depicted in Figure 10-4.

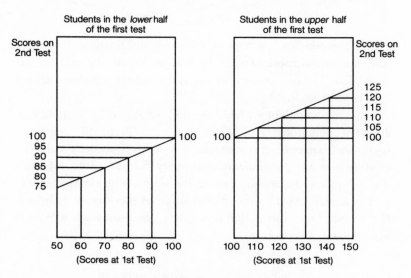

FIGURE 10-4
The Regression Fallacy: Seeing Only Half The Picture

Assume now that we look only at the right-hand half of Figure 10-4, at the students who in the first test performed above average, i.e., above 100. What will we see? The students who got a score of 150 on the first test averaged on the second test only 125, and all these students, to a different degree, will have scored on average somewhat lower on the second test than on the first one. Had we seen only the left half of the graph we would have concluded the opposite: on the average, all these students scored higher on the second test than they did on the first one. The trouble, of course, is that "all these students" are not really *all* students; they form a biased selection of all students.

Each of the two scores of every student consists of two parts: one reflecting the student's ability, the other reflecting a chance element that may vary from test to test and is unrelated to the student's ability. It is these latter variations that cause the regression effect.

HOW BIG A CHANGE?

The regression line, as Figure 10-5 shows, allows us to answer still another question, namely, by how much will the value of the dependent variable change as the independent variable changes by a certain amount?

If the independent variable moves, for instance, from 150 to 200, the dependent variable moves from 30 to 40. Since the regression behavior is summarized here in a straight line, the relation can be generalized: for every 50 measurement units the independent variable moves, the dependent variable moves on the average by 10 units; the relation of movements is 5 to 1. If the underlying correlation is negative, the movement will go in the opposite direction.

HOW MUCH IS EXPLAINED?

Having learned from the coefficient of correlation the degree to which two variables are correlated, we may now ask, "How much of the variation of the dependent variable is 'explained' by the correlated variation of the independent variable?" The precision

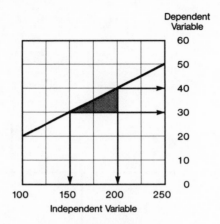

FIGURE 10-5
Interpreting the Slope of the Regression Line

with which we can estimate the size of the dependent variable from that of the independent variable will depend on the degree of scatter around the regression line, as we saw in Figures 10-2 and 10-3. We may now ask another question: *How much* of the variation of the dependent variable can be "explained" by the variation of the independent variable?

Figures 10-6 and 10-7 describe, with simple figures invented for this example, a real relationship, that between a city's homicide rate per 100,000 population and the size of that city.

Figure 10-6 records the homicide rates of 10 cities (A through J), with their average (mean) rate of 8.0 per 100,000. The deviations of the 10 city-rates from their average are what we want to explain, at least in part, by relating them to the respective size of their city.

The deviations are represented in Figure 10-6 by the lines that connect each city-rate with their common average of 8.0; Table 10-2 shows the respective numbers. Only the numbers in the first and second column are represented in Figure 10-6. In the third column, for reasons we need not go into here, the deviations are

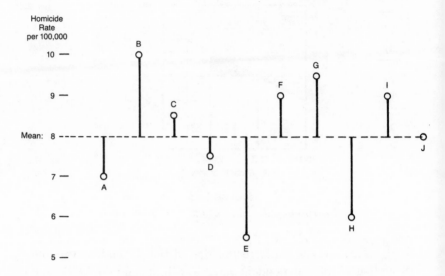

FIGURE 10-6
Homicide Rates per 100,000 in Ten Cities

squared, and the average of these squared deviations in our case is 2.0. That measure is called the *variance* and we will now learn from Figure 10-7 how one measures the amount of the variance that is explained through correlation, in our case by relating the homicide rates to the respective size of their city. In Figure 10-7 the ten cities are rearranged so that each is in its proper place on the horizontal axis. The upward pointing regression line denotes the positive relationship between homicide rate and city size.

To appreciate the mathematics of "how much is explained" we must understand the practical significance of the operation. If we were asked to predict the homicide rate for a city for which we had no data and knew only the data offered in Figure 10-6, the best we could do is to predict that its annual rate would be 8 per

TABLE 10-2
Deviations of the Homicide Rates from their Mean

City	Rate	Deviation from 8.0	Deviation Squared
A	7.0	−1.0	1.00
B	10.0	+2.0	4.00
C	8.5	+0.5	0.25
D	7.5	−0.5	0.25
E	5.5	−2.5	6.25
F	9.0	+1.0	1.00
G	9.5	+1.5	2.25
H	6.0	−2.0	4.00
I	9.0	+1.0	1.00
J	8.0	0	
Total	80.0	12.0	20.00
Mean	8.0	1.2	2.00
		percentage points	variance

100,000, the average of the ten cities for which we had data. We know that such a prediction has a considerable error range, on the average of ±1.2, and a variance of 2.0 percentage points.

If we knew in addition the size of the city for which we are to predict the homicide rate, we could make the prediction more accurate, as Figure 10-7 shows. There, in order to make the best prediction, we do not go up to the mean, as we did in Figure 10-6, but up to the regression line. The average deviation of the homicide rates for each city from the regression line, as can be seen, is smaller than the average deviation was in Figure 10-6. Table 10-3 compares the two variances (the average squared deviations) and thereby answers the question as to how much of the original variance was explained by the regression on city size.

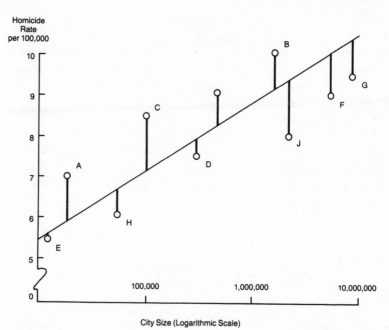

FIGURE 10-7
Relationship between Homicide Rate and City Size in Ten Cities

TABLE 10-3
How Much Did the Regression Explain?

	Percentage Points	
(a) Size of the variance to be explained (from Fig. 10-6 and Table 10-2).	2.00	(100%)
(b) The unexplained variance, the residuals which remain after the use of the city size regression	.82	(41%)
(a) − (b) The difference between the two, which is the explained part of the variance	1.18	(59%)

The city size explained 59 percent of the original variance; this
(a) − (b) value is called the *coefficient of determination;* its signed
square root is the familiar *coefficient of correlation, r;* hence the
coefficient of determination is r^2.

Table 10-4 shows the relationship between r and r^2. The
simplistic mathematics is spelled out here to guard the lay-user
of the correlation coefficient r against overestimating its explanatory power.

TABLE 10-4

Coefficient of Correlation (r)	Coefficient of Determination (r^2)
±1.0	1.00
±0.9	0.81
±0.7	0.49
±0.5	0.25
±0.3	0.09
±0.1	0.01

In the upper ranges the two coefficients are fairly close together,
but already a correlation coefficient of 0.7 indicates a reduction
of less than one-half the original deviations; and a coefficient of
0.3 corresponds to an r^2 of 0.09, a reduction of less than 10
percent of the original deviations.

MULTIPLE REGRESSION

Up to now we have considered only correlation and regression
between two variables in order to see the structure of that analysis
in its simplest form. The analysis, however, can be extended to
more than two variables, by our trying to find out how *several*
stimulus variables combine in their ability to predict or explain
the variation in the response variable.

If we want to know, for instance, how an agricultural yield is
affected simultaneously by the amount of applied fertilizer and

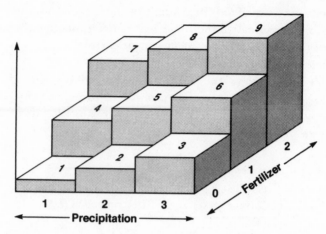

FIGURE 10-8
Yield under Various Conditions of Fertilization and Precipitation

the amount of precipitation, we must engage in multiple regression analysis. The results of such an analysis are schematically presented in the three-dimensional illustration of Figure 10-8.

Seen from the top, Figure 10-8 will look like Figure 10-9, the normal presentation of a three-variable relation. An increase in

Units of
Fertilizer

7	8	9	2
4	5	6	1
1	2	3	0

1   2   3
Units of Precipitation

FIGURE 10-9
Yield under Various Conditions of Fertilization and Precipitation—
Simplified View

precipitation by one unit increases the yield by one unit; as we move from left to right, each time adding one precipitation unit (the bottom scale), the yield is increased by 1 unit—irrespective of the level of fertilizer. As we move from bottom to top, each time adding one fertilizer unit (the right-hand scale) we increase the yield by three units—irrespective of the level of precipitation.

In this example, the two causal factors, precipitation and fertilizer, affect the yield additively; their combined effect is simply the sum of the two effects. For example, moving from 1 precipitation unit plus 0 fertilizer units (the cell on the lower left corner) to 2 precipitation units and 1 fertilizer unit, we gain 1 yield unit for the added one precipitation unit and 3 yield units for the added 1 unit of fertilizer. Hence the combination should contain $(1 + 1 + 3) = 5$ yield units, which it does in the cell to the northeast of the first cell.

It is of course possible and, as a rule, probable that certain combinations of precipitation and fertilizer will produce higher or lower yields because certain combinations will enhance (or reduce) their individual effectiveness. If there are more than three variables, the graphic illustration of Figure 10-9 is no longer available; algebra must take its place. But the principle remains the same.

THE PURPOSE OF REGRESSION ANALYSIS

The primary purpose of regression analysis is to predict the size of the dependent variable from that of the independent one. Sears Roebuck, for instance, estimated the daily dollar value of its incoming mail orders by simply weighing the mail. Sears found the weight of the mail to be a satisfactory early predictor of the approximate value and amounts of goods that had to be ready for shipment that day, thereby facilitating managerial arrangements. Orders on the books of tool manufacturers aid in one's predicting the general manufacturing level of the economy at a later point in time. Entrance examination scores help one predict school performance and thereby are a help for deciding which students should be admitted. Predictions can work even if there is no causal relation between the variables, so long as they are consistently correlated.

Often, however, regression analysis is used for drawing causal inferences, that is, for concluding that changes in the independent variable will *cause* certain changes in the dependent variable. Such inferences are relatively safe if they are based on controlled experimental data, but if they are based on observational data it is very difficult to draw safe inferences on cause and effect relations, although the outward form of the regression analysis is the same for both these applications.

Regression analysis of a controlled experiment, such as our hypothetical plant-growth example in Figure 10-9, will provide fairly precise knowledge of the effects that the two treatment factors produce. In that situation one proceeds from clearly defined stimulus (causal) variables and observes their effect on the response (effect) variable under controlled conditions.

CAUSAL ANALYSIS OF OBSERVATIONAL DATA

There are, to be sure, also observational situations where we would be fairly confident that the observed figures reflect a causal relationship. The data on the height of fathers and sons are an example, since the one clearly precedes the other in time and does not allow a reverse effect. But if we were to record for a sample of men, for instance, the amount of exercise they are engaged in and their state of health, the situation is not as clear. However well the one may allow us to predict the other, we could not be sure that exercise was the cause of good health. It could be the other way around; the healthier a man the more he will exercise.

In the regression analysis of observational data all the problems and dangers lurk which we began to discuss in Chapter 8 under the heading of true and spurious correlations. The difficulties begin with the need to locate and measure other variables suspected of affecting the performance of the dependent variable.

Suppose we wanted to determine the effect of speed on the fatality rate in automobile accidents. Even if we had reliable data for both variables at a number of places over several years, we would have to eliminate other influences on the fatality rate: smaller cars are more dangerous than larger ones, very young and very old drivers are a greater danger than drivers in the middle range, road and weather conditions must be considered, as well as drunken driving. All these influences may vary between places and over time.

Or suppose we wanted to find out whether the death penalty deters, by determining whether the frequency of executions affects the homicide rate. That rate, we know, is affected by a great many circumstances: by the number of persons, particularly men, in the crime-active years between 15 and 40, and by the economic, social, and cultural conditions of the area. To learn the effect, if any, of executions on the homicide rate we would have to separate their effect from that of these other influences.

Having determined the various factors that we must consider, the next problem arises: how to obtain reliable data for each of them. Data on highway fatalities are fairly good, but even there problems arise if the hospital in which a victim died is at a different place from the one at which she was injured. Driving speeds are now being monitored regularly on a sample basis, but the overall amount of driving, the denominator of the fatality rate, as a rule has to be estimated from the tax receipts on gasoline sales in the particular area.

Often we need some measure of the distribution of incomes. If no direct data are available we might have to rely on substitute data, for instance, on the occupation of the working members of the household. Since occupation and income are related one can try to estimate the latter from the former.

Thus the correctness of our causal inferences from the regression analysis will in the first instance depend on whether we have accounted for all competing causes and conditions, and whether we have reliable measurements for them.

TRAPS

In Chapters 8 and 9 we discussed at length the dangers of using correlation analysis for drawing inferences about cause-and-effect relations. We showed there how correlations may be spurious and thereby suggest false causal inferences. What applied there to the simplified categories of our examples applies also to the study of the fully developed variables in regression analysis. Figure 10-10 provides an example of how a scatter diagram may lead to a spurious inference about causation.[2]

The dots in panel A represent counties of varying population density denoted according to the horizontal scale at the bottom by their distance from the left-hand margin. The height of every dot denotes the automobile accident rate in these counties, as measured by the vertical scale. The oblong shape of the scatter diagram suggests an association between the two variables: the greater the population density in a county, the greater its rate of car accidents.

Upon examination, it turns out that the counties differ also in a third respect: some of them have obligatory car inspection and others don't (this is a hypothetical example). Panel B shows the counties that have such inspection (●); panel C includes the counties that do not have such an inspection (○). It turns out that the counties in the low-density area have inspection, presumably because it is a manageable operation there; the high density areas have no inspection. Neither of the two new scatter diagrams, both round and no longer oval, suggest any association between pop-

---

[2] The example was stimulated by another hypothetical example dealing with this topic in Edward R. Tufte, *Data Analysis for Politics and Policy* (Englewood, N.J.: Prentice-Hall 1974), p. 149.

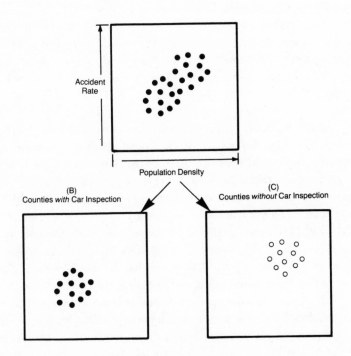

FIGURE 10-10
Population Density and Car Accident Rates in 29 Counties

ulation density and accidents. What apparently caused the difference in accident rates was the obligatory car inspection. The
combination of the two scattergrams in the oval shape of panel *A*
suggested a correlation that turned out to be spurious.

Another problem derives from the typical situation that some
of the independent variables are related to one another.

Next comes the problem that the direction of the causal
relationship between two factors may not be at all clear. The
difficulty is aggravated if the causal relation goes in both directions;
in some cases, *A* might cause *B* to vary; in other cases, *B* might

affect the variation of A. The correlation between income and education is an example. The better off a person is economically, the more extended, on the average, will be that person's education. In time, however, the longer education will be responsible for higher income. Or: growth of crime may lead to more police and more police may reduce crime. Only if the two movements can be sufficiently separated would their interaction become discoverable.

Recently, the news media put a study before the public which claimed to shake the generally accepted wisdom that low salt intake reduces blood pressure. The study showed that often high salt intake goes together with low blood pressure, and low salt intake with high blood pressure. A young friend, a healthy athletic young man has a very low blood pressure, and likes a great deal of salt with his food, partly perhaps to replenish the salt that evaporates during his hours of strenuous tennis. In contrast I, getting on in years, and for that reason having a somewhat higher blood pressure, do not indulge in eating salty food, although I, too, prefer the taste. Multiply my friend and me a thousand times and you obtain the statistics that allegedly confounds the theory that low salt intake reduces blood pressure. It does, of course, nothing of the kind, because the causal chain may be reversed: Low blood pressure encourages high salt intake, high blood pressure discourages it. The question not answered in that study is what effect a change in salt intake would have; how would my blood pressure react if I ate more salt, and how would my friend's blood pressure react, if in the process of getting older, he would eat less salt?

There is one form of regression analysis that must be treated with particular suspicion. Where correlation data for individual units are unavailable, one is tempted to use group averages—as a rule, geographically defined groups—as a substitute. If we want to learn, for instance, whether the amount of coffee drinking and the incidence of cancer are related, we want these data for a sample of individuals. In the absence of such data, investigators might look instead at different countries and see whether some countries show a correlation between per capita coffee consumption and the incidence of cancer. Depending on how the individual data combine to form the group averages, such *ecological* correlations, as they are called, may be quite unreliable, showing either higher or lower correlations than the individual measures would have produced.

Finally, there is the difficulty of deciding on the precise form, the "model," for each of the involved relations. The analyst has a wide choice of mathematical devices for describing an involved relation and thus has great freedom in the shaping of the regression formula. To give a simple example: It is often worthwhile to consider the possibility that the relation is logarithmic, so that a change in one variable produces a corresponding percent change in the other. All these regression formulations are by definition only models, and thus approximations of the real-world relations. How well a regression formula derived from observational data fits the real world is thus a delicate question. The formula's robustness, as statisticians call it, remains a pervasive problem; the general rule is to be wary of the results our computers spew out, often with deceptive apparent precision.

SUMMARY

Regression analysis is a powerful tool when it is used for the purpose of prediction. Dangers arise when we use regression analysis for making causal inferences about the effect of certain variables. The dangers are particularly great when the underlying data do not come from experiments but from observational data. In such situations the results must be accepted with caution, and efforts should be made to obtain from other sources independent corroboration of the observed regression effect, through "triangulation," as described in the concluding chapter of this book.[3]

---

[3] See entries CAUSATION and REGRESSION in the *International Encyclopedia of Statistics*, especially the postscript to the latter.

11

Reason Analysis I: The Accounting Scheme

THE ART OF ASKING, "WHY?"

In the preceding chapters we discussed the search for causal connections through the statistical analysis of relations between group measurements. This chapter turns our attention to a different method of discovering causal connections—tracing the causal chain of an *individual* event or action.

If the event we want to understand is a human action, appropriate conversation with the actor will be the best road to discovery. If the event is one in which human action played only a subordinate role, as in a factory explosion, or an event in which there were no survivors to talk to, as in an airplane accident, the search will have to take a more indirect route. That search for causes and reasons is always difficult. However, progress during the last decades has made it possible to sketch out here the beginnings of a technique of exploring individual events that has become known by the name of reason analysis.[1]

On first impulse, the question, "Why?" would seem to hold the key to all such endeavors. Where human actions are at issue, all we have to do, it would seem, is to ask the actor, "Why did you . . . ?" Strangely enough, that question as a rule yields

[1] See Paul F. Lazarsfeld, "Progress and Fad in Motivation Research," *Proceedings of the Third Annual Seminar on Social Science for Industry Motivation*, Stanford Research Institute, 1955, pp. 11–23. As to the praxis, see Herta Herzog's classic in Hadley Cantril, Herta Herzog, and H. Gaudet (eds.), *The Invasion from Mars—A Study in the Psychology of Panic* (Princeton, N.J.: Princeton University Press, 1940). See also Ernest Dichter, *Handbook of Consumer Motivation* (New York: McGraw-Hill, 1964); James V. McNeal (ed.), *Dimensions of Consumer Behavior* (New York: Appleton-Century-Crofts, 1965); Joseph Newman (ed.), *Unknowing the Consumer* (New York: Wiley, 1966); Karen A. Machover, *Personality Projection in the Drawing of the Human Figure* (Springfield, Ill.: Charles C Thomas, 1949).

disappointing insights. There are so many possible answers to the question that the interviewee often has a hard time giving a helpful answer. Suppose that we asked some recent immigrants why they had come to this country. Their answers would form a bewildering array of reasons. One will say, "We came because wages are low where we come from." Another, "Our uncle convinced us that it would be a good thing to immigrate." A third, "Because there are plenty of good jobs here in America"; a fourth, "My fiancé had come here some time ago, and I followed him."

It is not easy to see in what order one could put such a list of reasons, because they differ not only in details but also with respect to the dimension of the decision process to which they refer. The first immigrant spoke of dissatisfaction with the old country; the second, of a person who influenced him; the third, of the attractions of the new country; and the fourth, of a very personal constellation that decided the move. It is easy to see that all these statements, except the fourth, are at best incomplete. If poor wages in the old country were a reason to move, then better wages in the new one are the necessary correlate; and the influential uncle must have given some reason why he advised moving to the United States.

People generally have more than one reason for their actions, and the simple why question will fail to reveal them all, because we have a tendency to explain our behavior with one reason rather than with a long, involved story.[2] To be sure, the abbreviated one-reason answer may have a significance of its own: it may be the most important reason, a point about which we will have more to say later on. The girl who followed her fiancé is a good example.

Asking for reasons requires a more complicated apparatus, which generally involves a number of distinct steps.[3]

[2] When Willie Sutton, America's most famous bank robber, was asked why he robbed banks, he answered, "Because that's where the money is."

[3] See Paul F. Lazarsfeld and M. Rosenberg (eds.), *The Language of Social Research*, (New York: Free Press, 1955), pp. 387–391.

1. Formulating the problem to be explored, preferably in terms of the policy decisions which the findings are to inform.
2. Doing exploratory interviews and performing a data search.
3. Developing the accounting scheme (a term of art presently to be explained) for these types of actions.
4. Applying the accounting scheme to the survey operation.
5. Analyzing the findings.

We will deal with each of these steps in turn—the first three steps in this chapter and the remaining ones in Chapter 12.

FORMULATING THE PROBLEM

At first glance it would appear that when we set out to explore a person's motives, we should approach our task with an open mind, that is, without first constructing a framework in which the various reasons and causes can be accommodated.

A moment's reflection, however, tells us that a person's whole life history lies behind even the simplest choice and that his entire social and physical environment is implicated in every one of his decisions. Since it is neither possible nor desirable to pursue the network of reasons ad infinitum, the investigator must limit the inquiry. The boundaries of the frame will largely depend on the purpose of the inquiry. The first step, then, is to formulate this purpose and to decide on the range of relevant factors.

Suppose we had been charged by some government agency with exploring the causes of immigration, with finding out what makes the United States attractive to immigrants, and what obstacles there are to immigration. In such a situation, we would concentrate on all the "pulling" reasons of the move. Suppose, on the other hand, the old country had asked us to find out why so many of its families wanted to leave. In that case, our inquiry would primarily aim at the "pushing" reasons that drive the emigrants out.

Or suppose a crime commission wanted to learn how one could reduce the amount of burglaries in the community. Here one of the jobs would be to find out what leads men to become burglars: what factors in their background, their personal makeup, and the company they keep lead them to such a burglary or to a career of burglarizing. We would also investigate how burglars select targets and eliminate others as unsuitable; we should like to learn thereby what makes some premises more burglarproof than others. Finally, we will want to know what circumstances provide an effective deterrent to burglars.

The purpose of the inquiry will determine its emphasis and often lead to the important distinction between controllable causes and causes that are beyond control. In the crime study, for instance, it might be interesting but not very helpful to learn that youngsters from poor families are more likely to become criminals than youngsters from well-to-do families. But if it turned out that a youngster's attachment to a beloved teacher was what kept him straight, improving schools and teachers in the poor districts of our cities might become a move to be considered.

THE EXPLORATORY INTERVIEW

Many decisions have a relatively simple structure; with others we are sufficiently familiar, so that the eventual accounting scheme which is to accommodate the reasons for these decisions is easily constructed. But every so often the decision process we want to explore is more complicated than we think, or so unknown to the analyst that exploratory interviews are required as a first step in order to see clearly all the dimensions of the decision process.

If the data we seek are to come from interviews, we begin the process by conducting "pilot interviews" with persons who made the type of decision we want to learn more about.

Not much helpful guidance is available for the conduct of pilot interviews, only a few general rules. There is no harm in beginning the interview with, "Why have you . . . ?" and keeping the interviewee talking for as long as she cares. Later, it is often useful to discipline the interviewee's recall by challenging her to recount the decision process in its temporal sequence, exploring the influences and motives that had been operative at each decision point.

The task is a difficult one, because ideally it requires the combined skills of a psychologist, a historian, and an inquiring journalist, and the lawyer's ability to cross-examine.

In the main, there are three types of answers that require the interviewer to intervene with appropriate questions: when the answer is clearly insufficient, when the answer contradicts an earlier one, and when a link in the decision process is missing.

When the immigrant reports that her uncle persuaded her to come, we must ask what arguments he used. If a lawyer tells us that he has a rule never to waive his client's right to a jury trial and later tells us of a case in which he did, we must ask him to explain the exception. And if a convict tells us he burglarized a particular home because he is a professional burglar, we might want to know what made him decide on that home.

Since the interviewer lacks the legal authority of a judge, she must conquer through tact and empathy. Her task will often be made difficult because of the respondent's conscious or unconscious hesitancy to tell the truth.

A general interviewing rule is to recognize that the final choice is the endpoint of a funnel of successively narrowing alternatives. The interviewer traces back the reasons for which each alternative but the final one was eliminated. Conversely, any reason given for the final choice must preclude all other alternatives or be amended by additional reasons, which either singly or in combination do just that. At the end of a successful exploratory interview, the interviewer must be able to say to herself, "Now I understand."

From this type of preliminary investigation the structure, or the variety of structures, of the decision process will gradually emerge. Tracing the origins of burglaries, for instance, would quickly lead to the discovery that the decision process involved is of two kinds. The habitual burglar's last crime can be explained only by a much earlier decision to become a burglar. For others, the burglary may be a relatively isolated if not the first event of that sort. How people become professional burglars is one type of development; how they implement it by deciding on specific targets is another one. How a nonprofessional decides to commit a burglary will be a third type of decision process to explore.

Inquiries into the more harmless activity of moviegoing will encounter similar structural variety. Some moviegoers regularly visit their neighborhood theater, possibly not even caring what movie they will see. At the other extreme are moviegoers who want to see a particular film and will go to the other end of town to see it if necessary.

Our immigrants also probably fall into two groups: those who took part in the primary decision to come, and those who, like the fiancé or children, followed the decision of others.

Our preliminary inquiries will lead us also to the important distinction between individual and group decisions: a family may jointly weigh the arguments and decide to emigrate. The paradigm of a group decision, of course, is the trial jury, which by law is required to arrive at a joint decision.

Our insights into the variety of types of action will come partly from our accumulated knowledge and common sense, more often however from exploratory, informal interviewing. Once the research objective is formulated, this informal interviewing accompanies almost every decision step until the final research instrument is developed. The second step in our reason analysis consists of setting out the type or types of action we want to investigate.

DEVELOPING THE ACCOUNTING SCHEME

Once the multiplicity of reasons has been narrowed down to a manageable number, the next task is to map out, in preliminary fashion, the various dimensions of the process under study. By dimensions, we mean general categories of reasons, such as, in the immigration example, dissatisfaction with the old country, attractions of the new one, personal influences, etc. Or, in a study of automobile purchases, the circumstances that lead to wanting a new car in general, and the influences, such as recommendations and advertisements, which direct the potential purchaser toward a narrower choice of cars and eventually to the car that was bought.

The purpose of separating these dimensions is to encompass the variety of individual reasons within a logical frame so that eventually a more general statement about the reasons for making a decision can be made in statistical terms. What may seem to be idiosyncratic reasons must be grouped into appropriate categories.

A study designed to find out why women came to use a particular face cream elicited this typical set of disjointed reasons:

MISS A: I heard the cream advertised over the radio.
MISS B: I have very oily skin, and this cream is supposed to keep it dry.
MISS C: I have dry skin, and the druggist told me that this would keep it moist.
MISS D: It was supposed to have a pleasant smell.

Two things are obvious, just as they were in our immigration interviews: first, each of these respondents tells only part of the story we want to know; and second, the answers belong to different dimensions of reasons. Some refer to the quality of the cream; some, to how the respondent became acquainted with the product; and some, to the respondent's special needs for it.

In such an array of answers, we try to discern the structural dimensions relevant to our inquiry. From the face-cream example, the following three dimensions emerged:

1.   *The predisposition of the respondent*—special skin conditions or certain preferences or prejudices.

2. *The product*—its qualities, its supposed effects, its price, and so on.
3. *The source*—the means by which the respondent learned of the product or its qualities.

Preliminary exploration made it clear that a complete answer to the question of why she bought the cream would involve these three kinds of reasons, each referring to a different dimension. The accounting scheme in Table 11-1 allows us to see how complete or incomplete the answers of the four respondents were.

TABLE 11-1

Reasons for Selecting a Particular Face Cream

| Respondent | Product | | |
	Predisposition	Qualities	Information
Miss A	No answer	No answer	Radio
Miss B	Oily skin	Keeps skin dry	No answer
Miss C	Dry skin	Prevents dry skin	Druggist
Miss D	No answer	Pleasant scent	No answer

Table 11-1 is an accounting scheme of the simpler sort: respondent's predisposition, qualities of the product, source of information. The more general form of the first dimension, *predisposition*, includes all motives prior to the purchase decision; in one case a desire to have more beautiful hands, in another case, to be less lonesome. The second dimension refers to *attributes* of the desired object, and the third, to the *influences* which affect the course of the decision. Stated more generally, this set of three dimensions involves the person, the object, and the social setting.

This three-dimensional accounting scheme—in scientific parlance, a model—will fit many simple decisions in the marketplace and elsewhere.

We call it an accounting scheme because such a set of relevant and sufficient dimensions permits us adequately to structure the replies we receive for statistical bookkeeping purposes. It is an integration of the data into a model which both guides the collection of data and provides the framework for their interpretation. Without the generalizing device of the accounting scheme, it is difficult to analyze the diversity of individual motives. Reason analysis transforms the individual reasons for a decision into more general, quantitative statements about decisions of that particular type and, ultimately, about decisions in general.

THE PUSH-AND-PULL MODEL

Whenever a change or switch from one situation to another, or one product to another is at issue, a model recommends itself that is based on the two poles of the decision: the dissatisfactions with the old situation (push) and the satisfactions expected from the change (pull). The decision to migrate that we discussed at the beginning of this chapter is one instance; switching from one make of automobile to the next and selling the old house and buying a new one are other examples.

Around this polarity of push and pull, other dimensions will fall into place: the channels through which the new situation came to attention and the influences that strengthened pull or push, as the case may be.

Influences, in turn, may have two subdimensions: the *source* and the *content* of the communication. By relating the two, we are able to distinguish between the relative effectiveness of a message and of the source from which it emanates. At times, the source will be of no importance—for example, when we are told, "I ran home because I was told that somebody had broken into our house." At other times, it will be the source of the message that is important—for instance, when a boy tells us, "I did this because my father told me to."

The dimensions of the accounting scheme are the basis of the subsequent data search and of the eventual statistical analysis.

They must exhaust the kinds of expected reasons and they must be mutually exclusive, so that any reason can be identified as belonging to one dimension and no other. There is no limit to the scope of the accounting scheme; dimensions can be quite detailed or very broad; they may describe only a limited aspect of a decision or they may cover the entire process.

REASON ASSESSMENT

Reason analysis, as we pointed out, is not limited to interview data and conscious decision processes but applies also to other actions and events, such as the analysis of accidents. To distinguish that process from interviewing the actor, we must call it reason *assessment*. To be meaningful, the process requires that the analyst know something about possible cause-and-effect relations in that particular field of investigation; the assessment merely involves a decision about the causal sequence in the particular case. In short, the analyst must be an expert.

Investigators of traffic accidents have found the four-dimensional accounting scheme shown in Figure 11-1 to be useful.[4]

MULTIDIMENSIONAL MODELS

A rather complicated accounting scheme was required for the process by which defense lawyers decide whether to try their criminal case before a jury or—as they often have the right to do—to waive their right to a jury trial and have their client tried by a judge without a jury.[5]

We found two major types of such decisions: one, in which there was a standing rule in the lawyer's office either to waive or not to waive in a particular type of case. The other type of

[4] Stannard Baker, *Experimental Case Studies of Traffic Accidents*. Evanston: Northwestern University Traffic Institute, 1960. p. 5.

[5] The study was conducted as part of the Jury Project of the University of Chicago Law School, with the assistance of Howard Mann, then a student at the school.

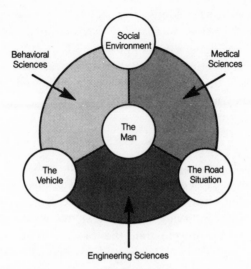

FIGURE 11-1
Accounting Scheme for Traffic Accidents

decision was made after more or less careful investigation of the
circumstances in the individual case. It is for this latter type of
decision that the following accounting scheme was developed.

Scheme of Reasons for Choosing Trial With or Without Jury

I. *Advantage Aimed At*
    A. More advantageous verdict
    B. Lower trial costs
    C. Better prospects for bargaining for an advantageous guilty
       plea[6]
    D. Better opportunities for appeal
    E. Better insulation against client if case is lost

---

[6] Under Anglo-American law, a defendant also can avoid trial altogether, by
pleading guilty to the charged crime, often to a lesser offense, possibly even to
an agreed-upon sentence. The vast majority of prosecutions end this way. See
Harry Kalven and Hans Zeisel, *The American Jury* (Chicago: The University of
Chicago Press, 1975), pp. 18ff. Also Zeisel, *The Limits of Law Enforcement*
(Chicago: The University of Chicago Press, 1983), pp. 34ff.

II. *Influences on Decision*
   A. Client's wishes
   B. Trial judge's preferences (to gain his favor)
   C. Counsel's personal preferences
   D. Countering opposing counsel's preference
      1. In particular case
      2. In general
   E. Tradition in the particular court
III. *Differences between judge and jury trial that can produce the advantage*
   A. Jury
      1. More than one person
         a. Composition can be modified through challenges before trial
         b. To convince one single juror could avoid conviction[7]
         c. Individual bias cancels out variety of views
      2. Basic attitudes
         a. Not always bound by letter of the law
         b. Specific bias
            (1) For underdog
            (2) Against unpopular indictment
            (3) Represents popular prejudice
   B. Judge
      1. More lenient on penalty if jury is waived
      2. Possibilities of personal bias
         a. Re: counsel
         b. Re: defendant or witness
         c. Re: type of case
IV. *The Case*
   A. Content of case
      1. Type of offense
      2. Is the primary issue a question of fact or of law?
      3. Defense effort concentrated primarily on question of
         a. Guilt
         b. Sentence
         c. Obtaining conviction for lesser offense
   B. Expected length of trial

[7] When the jury verdict must be unanimous, as in the federal courts and in capital trials in the state courts. Most other state court trials now allow majority verdicts.

    C.   Difficulty of case
    D.   Personalities in case
        1.   Client
            a.   Personal background
            b.   Physical characteristics
            c.   Past record
            d.   Manner of testifying
        2.   Expected witnesses
            a.   Personal background
            b.   Physical characteristics
            c.   Manner of testifying
            d.   Past record
    E.   Estimated odds of success (prior to trial)
    F.   Public attention received by case

A lawyer's reasoning for waiving a jury may then run as follows:

This was a case of a homosexual (IV, A, 1); I was somewhat afraid of a jury, because they don't like sex deviates [III, A, 2, b, (3)]. Also, I know the judge; he is an experienced wise man, not one of those hot rods (III, B, 2, c). There was anyway only a small chance of acquittal (IV, E); the big question was whether I could get a suspended sentence (IV, A, 3, b), and I knew that judges are easier on the sentence if they try the case without a jury (III, B, 1). On the whole, I thought I would get my client a better verdict without a jury (I, A). I talked to my client about this decision; he left it to me (II, A).

THE ART OF ASKING, "WHY NOT?"

At times the investigator wants to find out not what motivates action but what motivates inaction. The aim may be to strengthen these motives if the action is undesirable, as is the case with crime, or to counteract these negative motives if the action is desirable. An example of the latter situation are birth-control studies, particularly in underdeveloped countries, to determine why recommended birth control is not practiced. We reproduce, in Table 11-2, an accounting scheme for analyzing the reasons for that inaction.

TABLE 11-2

Accounting Scheme for Not Practicing Birth Control[*]

1. *Condition*

(i) Desirability of pregnancy   (ii) Inability-ability to conceive

2. *Knowledge*

(i) Of specific methods   (ii) Generally that there   (iii) Of neither
                               are methods

3. *Objections*

(To birth control in general and/or to specific methods)

(a) Substance   (b) Source
    (i) Morals   (i) Religion
    (ii) Health   (ii) Medical experts
    (iii) Comfort   (iii) Family, friends
                                         (iv) Prior experience

4. *Availability*

(i) Physically unavailable   (ii) Psychological obstacles

[*] Modified from David L. Sills, "The Art of Asking 'Why Not?' " in *Proceedings of the Fourth All-India Conference on Family Planning* (Hyderabad, Bombay: Family Planning Association of India, 1961), pp. 26, 33.

Since all four dimensions—need, knowledge, absence of objections, and availability—must become operative to motivate action, failure of one or all will impede it.

THE TIME DIMENSION

All decisions take place over a period of time. Even the most impulsive actions, such as an "impulse purchase" in a self-service store or a crime committed in a sudden flare of anger, can be meaningfully explained only in terms of preceding events. Time, therefore, is inherently a part of any decision process.

Not all accounting schemes, however, require a time dimension, because it is often irrelevant to learn the exact temporal sequence of a decision process. In purchase decisions of small items or in a lawyer's decision to try a case before a jury, the time dimension is of little interest. In other kinds of decisions, those that extend over a longer span of time, the time element will be an important factor. The purchase of an automobile or a major appliance, the choice of the candidate for whom one will vote on election day, a doctor's decision to begin prescribing a newly available drug require the inclusion of time into the accounting scheme.

PRECIPITATING EVENTS

There is a type of action evolving over time of which the end phase is triggered off by a relatively minor event. Latent motives for acting suddenly become effective when some external event or inner development activates the disposition and thereby forces the decision. The actions of the professional pickpocket may be imagined in this way. The experienced pickpocket is always disposed to ply his trade; he moves into action when some external event places him in advantageous proximity to a likely prospect.

There is also the other possibility that people may be impelled to act not only through the triggering effect of an external event but also by the force of some internal disposition that reaches action threshold. Crimes of passion can often be accounted for in this way. A seemingly inexplicable burst of violence may become understandable if we locate the pent-up emotions and the minor event that brought them to the surface.

This dynamic, if less dramatic, pattern emerged nicely from a study that explored why some people switched from coffee to tea.[8] For some people, there was a clearly defined external event, such as the sudden rise of coffee prices or doctor's orders to change. But for a significant number of people there was no such

[8] Phillip Ennis, *Why People Switch to Tea?*, Bureau of Applied Social Research for The Tea Council of the U.S.A., Inc. (New York: Columbia University Press, 1954).

visible event propelling the change. Dissatisfactions with coffee had simply accumulated to the point where "something had to be done about it," without any apparent trigger from the outside.[9]

PHASES OF DECISION

Then there are decisions that not only extend over time but change their character in the process. The process of acceptance of a newly developed drug by the medical profession seems to fit the basic predisposition–influences–attributes model, but with a difference. Three distinct phases appear to characterize the development. In the first, the *information* phase, the news is learned and absorbed; in the second phase, the physician is concerned with *evaluation;* and in the third, the *confirmation* phase, prior to accepting the drug, he looks for the experience of colleagues.[10] Table 11-3 shows how such a distinction of phases can provide insights into other aspects of the process. The changing influence

TABLE 11-3

Sources of Influence on the Physician at Different Phases of the Decision to Prescribe a New Drug

	Phase		
	I	II	III
	Information	Evaluation	Confirmation
Source	%	%	%
Pharmaceutical firms	80	56	45
Professional channels	20	44	55
	100	100	100

[9] The process during which the internal pressures accumulate is occasionally referred to as "crystallization," which is probably an unconscious echo of what is probably the finest example of an accounting scheme in literature: Stendhal's *De l'Amour*, in which the crucial and fatal phase of the falling-in-love process is named cristallisation.

[10] E. Katz, P. F. Lazarsfeld, *Personal Influence* (Glencoe, Ill.: Free Press, 1963).

of channels, pharmaceutical firms at first and later professional sources, suggests that the channels have different functions in different phases of the process.

Similarly, occupational choice decisions are fitted best into an accounting scheme of three decision phases.[11] The first, the *fantasy* phase, spanning roughly the years from 6 to 11, serves to determine the range of available alternatives which the child explores according to the pleasure they promise. The second phase of *tentative choice*, roughly the years between 11 and 17, determines the realistic range of alternatives. In the last phase, the *selection of the specific occupation*, a variety of influences and considerations precipitate the final choice.

NARROWING THE CHOICE

A special case of this multiphase decision is a process that moves from broad general delineations to ever narrower ones and suggests an accounting scheme of decisions of ever-increasing specificity. This model is particularly pertinent when the original number of choices is large. Disk jockeys, for instance, select the music for their broadcasts from literally hundreds of thousands of available records. Yet the disk jockey makes his selection in a relatively short time. How does he do it? The process begins with the decision to be a particular type of disk jockey. He can become an *air salesman*, emphasizing that part of his work which deals with commercial sponsors; he can become a *music promoter*, emphasizing his responsibilities as a popularizer or even creator of hit records; or he can become a *broadcast personality*, emphasizing his job as entertainer. The decision in turn determines the audiences to which he must appeal. The air salesman must try to reach a buying public; the music promotor must appeal to the mostly young music enthusiasts; and the radio personality, through his program, must build from a yet undifferentiated audience.

---

[11] Eli Ginzberg et al., *Occupational Choice, An Approach to a General Theory* (New York: Columbia University Press, 1951).

In the next decision phase he must select the *type* of music he will play, a choice largely determined by the previous decision. The presumed musical tastes of his audience dictate the proportion of old favorites, top hits, and new records.

Another, more common, accounting scheme of narrowing choices is standard for the analysis of purchases of major appliances and automobiles. Here the general decision to buy is narrowed to a particular make, to a particular model, to a particular color combination, and so forth. The sequence of these decisions will vary; the decision to buy another Buick may be made long before the purchase of a new car is decided upon; the decision to buy a sports car may be the very first one in the chain of narrowing choices.

SUMMARY

Within limits, appropriate questioning of the actors will elicit the motives for their actions. A necessary step in this process is the development of an accounting scheme, a model of the action to be explored. The scheme serves as a guide both to proper interviewing and to subsequent statistical analysis of the data; and it is developed after preliminary, informal interviewing. The accounting scheme must take care both of the different time phases of an action and of the multiplicity of influences at any one time: it does this by establishing different *dimensions* within the action model. Many simple actions, especially purchase acts, will be adequately covered by the three-dimensional scheme that allows for motives that originate in the actor, for influences from his surroundings, and for the properties of the object in question. More complicated accounting schemes are also discussed.

12

Reason Analysis II: Data Collection and Interpretation

ACCOUNTING SCHEME AND QUESTIONNAIRE

Once the accounting scheme is established, the data-gathering process begins. At that point, if all has been done correctly, one knows the types of data that will fill out the categories of the accounting scheme. What one does not know are the more specific reasons in each category and, in particular, their frequencies. It is well to remember that the accounting scheme is only a logical structure designed to accommodate any set of actual reasons. The scheme is thus more complete than any real explanation pattern, but also less concrete than the psychological reality of actual behavior.

Unless its dimensions can easily be translated into questions, the accounting scheme must not be used in lieu of a questionnaire. It would form so specific and rigid a schedule that it would force the respondent to think in the interviewer's terms instead of his own. On the other hand, if the interview proceeds entirely by loosely knit, open-ended questions, it might fail to cover all the relevant dimensions.

The desirable solution is to write the questionnaire around the categories of the accounting scheme, allowing at the same time sufficient freedom to the interviewee to tell his story in its natural sequence.

The following questionnaire, which was aimed at finding out why students chose a particular college, may serve as example. It is quite proper to begin with the general question, "Why?", provided it is followed up by supplementary inquiries:

1. When planning your college years, why did you decide on ——— college?
2. *(Supplementary questions. Ask only those which were not answered in 1.)*
    a. What notions of expected qualities of the college influenced your decision?
    b. How did you happen to know about these qualities of the college?
    c. What were your particular interests or needs in the choice of a college? Which of your own particular needs seemed to be taken care of?
    d. Did other persons influence your decision to go to this college? Who were they? What did they tell you?

Some respondents will give a complete answer to question 1; from others, only the supplementary questions will elicit the total picture. Whether this is a difference of articulateness in recalling the decision or a true difference in emphasis is one of the problems we will return to later on.

PROBING

The major interviewing problem derives from the difference between what the respondent thinks is a satisfactory answer and what the interviewer regards as satisfactory. The interviewer must be familiar with the accounting scheme because she must try to obtain answers for each of the dimensions.

To this purpose, the interviewer must help her respondent along by encouraging more specific or more complete answers, or by asking for the resolution of contradictions. A lawyer may tell us that he insisted on a trial by jury in a particular case "because I always prefer a jury." If upon specific questioning, he then recalls a case where he did waive his right to jury trial, the interviewer will be entitled to an explanation. If the respondent wilfully or unconsciously hides elements of the true answer, the interviewer's probing may approach the lawyer's art of cross examination.

A good many millions of people have watched one of the best

probing interviews of this kind when they saw the lovely Katharine Hepburn as Jane in the motion picture *Summertime* quizzed by her admirer, the Venetian antique dealer:

JANE.   Signor de Rossi . . . why did you come to see me?

DE ROSSI.   It is only natural. You are not going to keep buying glasses every day.

JANE.   No.

DE ROSSI.   So I came.

JANE.   But why?

DE ROSSI.   Listen—two nights ago I am in Piazza San Marco . . . you are in Piazza San Marco. We look. Next day you are in my shop. We talk about glasses—we talk about Venice— but we are not speaking about them, are we? No. So last night I am in Piazza San Marco again. You are in Piazza San Marco again.

JANE.   Half of Venice is in Piazza San Marco again.

DE ROSSI.   But half of Venice was not in my shop this afternoon, or I would be a rich man.

JANE.   I wanted to buy another glass.

DE ROSSI.   That's all? . . . There are shops all over Venice. Did you look in any of them for your glass?

JANE.   No.

DE ROSSI.   You see?

JANE.   But you said that you would find one for me.

DE ROSSI.   And that's why you came back.

JANE.   Yes.

DE ROSSI.   No other reason.

JANE.   Signor de Rossi, I'm not a child, but I don't understand.

DE ROSSI.   Understand? Why must you understand? The most beautiful things in life are those we do not understand. When we spoke yesterday, I knew you were simpatica. Is that something you understand?

JANE.   Yes. It means I am like a sister to you.

DE ROSSI. Miss Hudson—you ask me why I came here to see
   you.

The general rule for probing is to recognize that the final
choice is the endpoint of a funnel of successively narrowing
alternatives. The interviewer traces back the reasons for which
each alternative except the final one was eliminated. Conversely,
any reason given for the final choice must preclude all other
alternatives or be amended by additional reasons which, either
singly or in combination, do so.

VERIFYING ANSWERS

Memory fades with time and tends to be distorted by intervening
events. Such errors may surface during the interview in implausible
or even contradictory answers. Also, without such warning signs,
the interviewer, wherever possible, should seek corroboration by
appropriate cross questioning, occasionally even by an independent
search for objective data.

Verification is especially needed when people report on such
difficult questions as their exposure to various channels of information. People tend to underestimate, for instance, their own
exposure to advertising and are reluctant to admit its influence.
We cannot force the respondent to recall nor can we force him
to admit. However, judicious questioning and requests of "play-
back" may reveal the extent to which the vocabulary of a
particular advertisement or series of commercials has been absorbed.

For influences of high prestige, the opposite is often true.
People, for instance, tend to overstate the number of serious
books they have recently read. In such situations it will help to
insist on details that are difficult to invent.

The proffered incomplete reason will sometimes serve as a
signpost for the unexpressed reason. In the beverage study mentioned in Chapter 11, it was found that people expressed their

reasons for giving up coffee in conventional and commonplace terms such as, "It upset my stomach," or "It was too stimulating." These replies formed the foreground to more emotionally laden attitudes of hostility to coffee which, though not stated as reasons, provided the latent impetus for giving up coffee. Because of the stimulating power attributed to coffee, some respondents blame or at least associate their coffee drinking with their misbehavior in social situations. Identifying such connections with psychological layers that do not lie on the surface is one of the primary functions of reason analysis.

HOW FAR TO SEARCH?

If a press report reads, "The Missouri river reached flood stage because upstream yesterday 5 inches of rain fell," we will feel that we understand. There is no need for further explanation. But suppose that the effort to understand the causes of the flood is part of an official inquiry about why the residents on the riverbank had not been warned in time. In such a situation, we would want to know more about the 5 inches of rain and the aftermath. Five inches over how large a territory? Five inches into half a mile of the riverbed is not the same as 5 inches into 2 miles. How concentrated was the rain? Five inches within 2 minutes or within 2 hours? How much time elapsed between the rainfall and the flood downstream? Did the authorities know the amount of rain, and if not, should they have known it, and so forth.

Other situations will not demand such lateral extension of the explanation, but may demand a search for further links in the causal chain—the reasons of the reasons. If human actions are at issue, every causative condition theoretically can be linked to a prior condition in a widening chain of causation that would end in a complete biography of the actor and a history of his environment. How far back we should trace the chain of reasons will depend on the purpose of our inquiry, its diminishing yield the further back we go, and its increasing costs.

The analyst might care to stop his search for causes simply because he is not interested in further pursuit of the causal chain. If in a survey of flower buying, Mr. X tells us that the last time he bought flowers was on his wife's birthday, we may not be interested in exploring why he did this—because of having love for her and of flowers, or fulfilling a routine performance that his secretary had to remind him of, or whether out of a bad conscience. Such variations may be interesting in another context, but not in this one.

Sometimes, it is only the more remote link in a causal chain that will require explanation. A while ago, the following item in *The New York Times* aroused some merriment among its readers: "There was an increase of 525 in the number of unmarried, pregnant students in the New York schools over 1961–62. The Bureau of Attendance had no explanation for the increase." It is all a matter of the right accounting scheme.

Occasionally, we will *want* to pursue a causal nexus further but the respondent is unable to help us, despite our asking her all the proper questions. This is the point at which the technique of *reason analysis* reaches its natural barrier, and what is loosely called *motivation research* begins. This will occur, for instance, when a reason has been pushed back to the level of individual taste: "Why did you prefer the blue car to the green one?" "Because I like blue better."[1]

To take a less-frivolous example: How can one discover the psychological effects of long-lasting unemployment? Efforts to ask, "What did unemployment do to you?" will produce only the fringe aspects of a complex phenomenon.[2]

This is not the place to discuss these difficulties further; only the general strategy of overcoming such research barriers may be

---

[1] And the question, "Why did you marry Jean?" will at best yield meaningless banalities.

[2] See Marie Jahoda, Paul F. Lazarsfeld, Hans Zeisel *Marienthal, The Sociography of an Unemployed Community* (Aldine-Atherton, Chicago, 1971). This study of an Austrian village done in 1930 was first published in 1932 by Hirzel in Leipzig.

indicated. Instead of expecting the respondent to give more explanations, the interviewer must develop and test his own hypotheses about causal connections which the respondent may not be aware of.

To begin with a color example, the choice of color for an automobile. It is likely to be determined by two factors: personality traits that have affinities to particular color preferences and fashion trends. Psychological tests and inquiries into the creation of fashions will be the appropriate research tools.

The search for the effects of long-lasting unemployment involves a variety of approaches. A school paper, "My Christmas Wishes," revealed that the children of unemployed workers wished much more modestly than did those of employed workers. And the school paper, "What I want to be one day," showed that occupational ambitions also shrank under the burden of unemployment. Observation in the streets revealed slower locomotion as part of the broad and deep effects of having lost one's job.

Sometimes the approach is limited to or aided by standardized tests such as the Minnesota Multiphase Personality inventory or the Rorschach test. More often it is a test developed for the specific purpose that asks the interviewee to answer a set of attitude questions or to draw a picture as best as he can of a specified object. The usefulness of all such efforts is predicated on the analyst's ability to establish a meaningful connection between the revealed attitudes and the behavior that is to be explained.

PRIMARY AND SECONDARY REASONS

Any action or for that matter any event is the result of the confluence of a great many causes; seldom, however, will we be inclined to attribute equal weight to all. Some of these causes, or even only one of them, may be singled out as more important than the others.

What makes some causes more important than others is determined, as we shall see, in part by the context of our inquiry. Following are examples of singling out the most important reason: "I came to the United States because my husband decided to go there and I followed him;" "this car had an accident because it was driven at excessive speed;" "X died because he was murdered."

Obtaining a variety of reasons, all pointing to the same result, reassures the interviewer. But in another respect, this variety is unsatisfactory. If somebody acted on the advice of several people, one would like to know whether some of them were more influential than the others. One would like to know the relative importance of the reasons given, first within the same dimension of the accounting scheme. If the reasons fall into several dimensions, one would like to know whether one dimension stood out in importance. If we are told, "We went to the neighborhood movie theater because my girlfriend wanted to go there, and because I read good things about that particular film," we would like to know whether the good things were as important a reason as his girl's wish.

The assessment of such relative importance may come from the actor-respondent himself, an observer, or the analyst. The actor may be requested to rank the given reasons according to their importance, first within the same dimension of the accounting scheme; afterward he might be asked to rank the dimensions themselves. He might be asked whether if reason X (and in turn reason Y) had not existed, would he have nevertheless acted as he did. If the truthful answer is yes to reason X and no to reason Y, the approach was successful: reason Y was the important one. Often, however, it will be either no or yes to both questions, which would not necessarily prove that both reasons were equally strong or equally weak; degrees of importance may matter.

The boy who went to the movies because his girlfriend wanted to and because he wanted to see the particular film would conceivably admit that had he known nothing about the film, he would have gone anyway and thus reacted to the most discriminating test question. Failing to obtain the admission, the analyst himself might come to that conclusion after probing. He or an observer may often do better in evaluating the relative importance of reasons than the actor himself.

Sometimes the very sequence of the narration will reveal something about the relevance of reasons. On issues that are discussed without embarrassment, the first reason given may be the most important one; on touchy issues, the reverse might be more probable.

At this point it will be useful to delineate our problem more precisely, both negatively and positively. If each of several reasons was a necessary condition of a completed action, it is not clear how one reason could have been more important than the others.

Relative importance might nevertheless emerge, as two examples will illustrate. If we ask why the emperor of Austria declared war in 1914, we cannot ignore the murder of the crown prince in Sarajevo. Still, historians do not believe that this assassination was as important as were other broader political issues. What they mean, of course, is that the situation was such that a variety of other small and rather probable events might well have triggered the declaration of war.

A reason might then be accorded relatively little importance because of two circumstances: it must be something like the straw that broke the camel's back; and the likelihood must be great that if a particular event had not occurred, another incident would have taken its place in essentially the same chain of events.

A second illustration leads to a somewhat related point. It is taken from the already mentioned study of the American jury. Sometimes a jury arrives at a verdict that differs from what the presiding judge thought was the correct one. With respect to such a situation, the judge told us why, in his opinion, the jury's views differed from his. One judge told us, for instance, that in a case before him he would have found the defendant, who had killed her husband, guilty of murder, whereas the jury convicted her only of manslaughter. He gave two reasons for the jury's

TABLE 12-1

Relative Importance of Two Reasons for Acquitting a Defendant of the Charge of Murder

(Figures are fictitious)

	Motive of Homicide	
Defendant	Justified Jealousy	Other Motives
Had a small child	60 percent acquittals	30 percent acquittals
Had no small child	50 percent acquittals	20 percent acquittals

leniency: The victim had openly consorted with another woman and the defendant had a small child. He felt that the first reason was the more important one. This assessment could have been based on a conversation the judge had with the jurors. If he more probably had no such conversation, he was guided by his experience with juries in similar situations, which could be formalized by the following fourfold table, Table 12-1. It gives the percentage of findings of guilty of a lesser offense than murder for the four types of cases.

The cases in the upper left-hand cell, which represent homicide motivated by justified jealousy by a defendant who had a small child, resulted in 60 percent acquittals. The acquittal rate for the cases where the spouse had no reason for jealousy and no small child was only 20 percent; and so on. Having a child increases the percentage of acquittals by 10 percent (from 50 percent to 60 percent, and from 20 percent to 30 percent); but jealousy because of adultery increases the percentage of acquittals each time by 30 percent (from 30 percent to 60 percent, and from 20 percent to 50 percent). Not the numbers but notions of such possibilities may induce the judge to consider jealousy the more important reason of the two.

Intuitively, there would seem to be a counter argument against the possibility of distinguishing more or less important parts, where all are needed to sustain the whole. The simile of an automobile motor comes to mind. In what sense can one meaningfully say that the crankshaft is more important than the cable that connects the battery with the starter? Obviously both are equally essential to the motor's performance. But if we think of how much time and money are needed to replace the failing part, we can arrive at a meaningful yardstick of "importance." It is on considerations such as these that an assessment can be made in an individual case. That such assessment in practice will prove to be difficult is another matter.

A corollary advantage of separating the important reasons is that it simplifies cross tabulation and multivariate analysis. If each action can be classified under one, and only one, category, or at least under no more than two, the analytic operations are facilitated.[3]

The problem about which among the causes of an event was the more important transcends the report of the actor. Consider an automobile accident in which the driver was drunk and drove at excessive speed, both of which situations are known to increase the likelihood of an accident. Is a meaningful statement possible about their relative importance? If the car had veered in the opposing traffic lane, I suppose one could say that drinking was the more important cause, since in a head-on car collision, speed is a minor issue; any speed is likely to be fatal. If no such individualization is possible, a cross tabulation of the sort presented in Table 12-1 would be helpful. It would show the effect on the accident rate of the four combinations of drunk vs. sober and excessive vs. normal speed.

[3] William H. Kruskal, "Concepts of Relative Importance," *Qüestió*, Barcelona, vol. 8, 1984, p. 39ff.

SUMMARY

Before the accounting scheme is fully developed, some preliminary questions must be answered. How far into the past or how deeply into the unconscious the search should go will depend on the purpose of the survey and its budget. Since every action has more than one cause or reason, it becomes desirable to distinguish more from less important reasons. A suggestion is made about how this complicated task might be approached.

13

The Panel

OVERVIEW

A group of individuals, households, or other units that serves as a continuous source of information is what social scientists call a panel.

The panel grew out of the needs which the one-shot survey could not meet or could meet only poorly. Observations of individual behavior over time, especially time of growth of youngsters (Pestalozzi) or infants (Piaget) has an old history; the panel is a fairly recent invention.[1]

The method of selecting a panel is not different from that of establishing its survey counterpart. In both cases it is a sample designed to represent a universe—of voters, households, doctors, grocery stores, or simply the general population. The distinctive characteristic of the panel is its relative[2] permanence as a source of information at successive points of time. The major advantage it offers over the one-shot survey is the opportunity to observe decision processes while they occur, thereby allowing them to be seen in slow motion, with precision and detail.

Efforts to reconstruct a decision process through a one-shot survey are hampered by two limitations: by the natural time limits of an interview and, more importantly, by the limits of memory. Some events stay with us for a long time, at least in outline. We

[1] The first systematic presentation of this technique was published in 1938 in a paper by Paul F. Lazarsfeld and Marjorie Fiske, "The Panel as a New Tool for Measuring Opinion," *Public Opinion Quarterly*, vol. 2, 1938, pp. 596–612. The novelty of the technique concerns the observation and quantitative analysis of group behavior. Observations of *individual* development over time have long been practiced; first, of course, in botany and zoology; later in pedagogy (Pestalozzi), and more recently in genetics.

Probably the oldest systematic effort to relate group observations over time resulted in the first mortality statistics by John Graunt in 1662. To be sure, these first observations were made retrospectively from birth and death registries.

[2] Relative, because as time passes, panel members drop out and are replaced.

all remember the schools we went to, or the two last automobiles we owned, or for whom we voted in the last election. Most events, however, are quickly forgotten; the memory of a minor expenditure or of what we read in the newspaper or saw on television will disappear within days. Thoughts and reactions might disappear even quicker. Moreover, to rely on memory is always a treacherous undertaking because of the unconscious forces that tend to distort it.

A particularly frequent memory error concerns the placement of an event in time. Questions such as, "Did you buy a new automobile during the past 12 months?" will invariably yield numbers in excess of control figures, because such a pleasurable and important event is consciously or unconsciously squeezed— "telescoped" is the technical term—into the response, even though it occurred 13 or 14 months before the interview.

Even if the event itself is correctly remembered, the subtleties of motivation and details of the causal chain that led to it are as a rule lost to the investigator, unless they are recorded while they occur or shortly thereafter. If we want to find out how a voting intention crystallizes or changes during a campaign, how a television viewing pattern develops, or how a purchasing pattern is related to brand loyalty, the one-shot interview will seldom suffice. Only a panel will provide the full information.

The panel operation allows repeated contacts and therefore potentially yields more information. This is true not only for the amount on current events but also for the storage of background characteristics that will grow in size after each interview.

Contact with the panel members may take a variety of forms. The initiation into the panel and the credible assurance of future cooperation form the important first step. Since the value of the panel depends on the quality of the statistical sample it represents, the initiation of the designated panel member must not fail. Just what is needed to secure cooperation should be explored through pilot tests, that is, operations in which incentives for cooperation are tested. Much will depend on how much cooperation is required from the panel. The burden may vary sharply between such relatively pleasurable activities as being periodically interviewed about some interesting episodes, and such onerous activities as keeping a continuous diary of household purchases or of television viewing for the whole family.

Since personal door-to-door interviews are costly, good strategy will aim at accomplishing as much as possible over the telephone or through the mail, or through a combination of these less expensive and often superior means of communication.

CONCEPTS COVERING TIME

The panel approach will be particularly valuable for the accurate establishing of research concepts that have a time dimension. We speak loosely of a *regular listener* or an *occasional user;* in ordinary conversation such terms may suffice. In a scientific study, they require definition.

To rely on the respondent's memory for the pattern of using or listening involves the danger of bias. To let the respondent herself decide whether she is a regular listener or user is even worse, because we will never know how that translates into numbers and how that translation varies from one respondent to the other. Exactly where to draw the line between a *regular* and *occasional* user or listener will in any event involve some arbitrary decision, but the panel can provide at least an exact record of how often and in what intervals the product has in fact been used, or a program been listened to—a solid data base for classification.

SHIFTS AND CHANGES

One of the points that sharply demonstrates the superiority of the panel operation is the analysis of shifts. As an example we will use the well-known polls which at intervals assess the president's popularity with the voters.

Table 13-1 records the results of two Gallup polls on the

TABLE 13-1

Shift in Approval of Former President Johnson's Policies 1965–1966

	January 1965 %	January 1966 %
Approve	74	58
Do not approve or undecided	26	42
Total	100	100
(Number of interviews)	(1,500)	(1,500)

popularity of former President Lyndon Johnson—one conducted in January 1965 and the other a year later.

We learn from Table 13-1 that during the year 1965 the proportion of people who approved the administration's policies dropped from 74 to 58 percent. No other aspect of this change can be learned from two such polls, except, of course, the comparable figures for subsamples, such as the proportion approving among Democrats, Republicans, blacks, women, voters under 30 years, etc.

Suppose now that instead of two successive polls, a panel had been established and the respondents of the first poll were reinterviewed a year later. In that case we would be able not only to learn the net balance of the shift but also to know *for every individual* where it stood at the time of the first and the second interview; a table such as Table 13-2 could then be constructed.

Table 13-2 presents a cross tabulation of the 1965 poll of the panel against the poll of 1966. The marginal line at the bottom reproduces the overall result of the 1965 poll; the right-hand marginal column gives the result of the corresponding 1966 poll.

The new information provided by the panel approach, that is, by interviewing the same respondents in both years, is contained in the four cells that form the core of Table 13-2. Since no such panel was interviewed, I have invented the four figures in Table 13-2 and Table 13-3 to illustrate the power of the panel.

TABLE 13-2

Shift in Approval of President Johnson's Policies, 1965–1966

Hypothetical Case A

1966	1965		Total 1966
	Approved %	Did not approve or undecided %	%
Approved	58	. . .	58
Did not approve or undec.	16	26	42
Total 1965	74	26	100
			(1,500 interviews)

The upper-left-hand cell contains the 58 percent of all respondents who approved of Johnson in 1965 and still approved in 1966; the lower-right-hand cell contains the 26 percent who disapproved in 1965 and still disapproved in 1966. The only group that changed was the 16 percent who approved in 1965 and disapproved in 1966; they account for the 16-point decline in approval from 74 to 58 percent.

The marginals of Table 13-3 are again the same as in Tables 13-1 and 13-2, but its core is quite different. Here all the 26 percent respondents who disapproved of Johnson in 1965 switched to approval in 1966; but a majority of the 74 percent who approved of Johnson in 1965, namely, 42 percent of all respondents, shifted from approval to disapproval. The difference between these two shifts, 42 percent away from Johnson as against 26 toward Johnson, produced the net decline of 16 percentage points.

Tables 13-2 and 13-3 were laid out so as to represent the

TABLE 13-3

Shift in Approval of President Johnson's Policies 1965–1966

Hypothetical Case B

1966	1965		Total 1966
	Approved %	Did not approve or undecided %	
Approved	32	26	58
Did not approve or undec.	42	. . .	42
Total 1965	74	26	100 (1,500 interviews)

extremes of the internal shift patterns compatible with a marginal change from 74 to 58 percent. Since extreme constellations are always improbable, the actual shift pattern surely lies somewhere in between.

TURNOVER AND NET CHANGE

Table 13-4 shows the basic structure of the fourfold switch table.

The upper-left-hand (Yes → Yes) and the lower-right-hand cells (No → No) contain the respondents who did *not* change their position; the two cells along the other diagonal contain the panel members who did change their position. The words yes and no stand, of course, for any dichotomy such as "approve–don't approve" or "use–don't use."

The two cells in the heavier frame (2) and (3), representing those who switched their position, can be combined so as to develop two crucial measures, turnover and net shift. *Turnover* is defined as the proportion of respondents who shift their position,

TABLE 13-4
The Basic Shift Table

Second Interview	First Interview		
	Yes	No	
Yes	Yes → Yes (1)	No → Yes (2)	Total *Yes* at 2nd interview (1 + 2)
No	Yes → No (3)	No → No (4)	Total *No* at 2nd interview (3 + 4)
	Total *Yes* at 1st interview (1 + 3)	Total *No* at 1st interview (2 + 4)	100% (1 + 2 + 3 + 4)

the *sum* of cells (2) and (3), expressed as percentage of all panel members:

$$\text{Turnover} = \frac{(2) + (3)}{(1) + (2) + (3) + (4)}$$

The turnover in Table 13-3 was 240/1,500, or 14 percent. It was (630 + 390)/1,500, or 68 percent, in Table 13-4.

The *net shift* is defined as the *difference* between cells (2) and (3), again expressed as a percentage of all panel members:

$$\text{Net shift} = \frac{(2) - (3)}{(1) + (2) + (3) + (4)}$$

The net shift in Tables 13-3 and 13-4 is, of course, the same, namely, −16 percent. Note that the net shift can be either positive or negative, depending on whether the figure in cell (3) is smaller or larger than that in (2). Note also that while we have shown

TABLE 13-5

Percent of United States Families with Checking or Savings Accounts of the Indicated Size in 1964 and 1965

1965	1964				Total 1965
	None	$1–1,999	$2,000–4,999	$5,000 and over	
None	14	6	1	°	21
$1–1,999	5	29	5	2	41
$2,000–4,999	°	4	5	4	13
$5,000 and over	°	3	4	18	25
Total 1964	19	42	15	24	100% (all families)

° Less than 1/2 percent.

here how the net shift can be developed from panel data, it is a measure for which the panel is not needed; successive one-shot surveys produce it too.

MULTIPLE SHIFTS

Turnover and shift figures need not be limited to simple two-by-two tables. They can be more complex as, for instance, in Table 13-5, which shows for two points in time—summer 1964 and a year later—the amount of money held in bank accounts or United States government bonds by a probability sample of United States families.

The bottom line shows the 1964 distribution of families by the size of their liquid assets; the column at the right-hand margin shows that distribution for 1965. The overall distribution remained fairly stable: in no group did the percentages change by more than two points: 19 vs. 21, 42 vs. 41, 15 vs. 13, 24 vs. 25. The core of the table, however, reveals great shifts. The diagonal line connects the proportion of families who stayed in their bracket, altogether 66 percent. The remaining one-third of the families shifted to a different bracket; the 18 percent above the diagonal shifted downward; over 7 percent (6 + 1 + *), for instance, had some assets in 1964 and none in 1965. The proportions below the diagonal designate the families who shifted upward; for instance, over 5 percent (5 + * + *) had no assets in 1964 but some in 1965.

SHIFTS AT SEVERAL POINTS IN TIME

So far we have dealt with shift at two points in time. If the number of time points is increased, the shift picture becomes more complicated. Take as an example the shifts arising from three panel interviews designed to ascertain the voters' approval or disapproval of the administration's policy during the war with Japan. Two alternatives at three different points of time make for the eight possible shift patterns shown in Table 13-6, where the plus sign stands for approval and the minus sign, for disapproval. The percent figures in the last column reflect the pattern found in a panel interviewed by the Gallup poll during World War II.

TABLE 13-6

The Eight Shift Patterns for Three Successive Interviews
Concerning Approval (+) or Disapproval (−) of the
Government's Conduct of the War

Shift Pattern	Interview			Percent of Panel Members
	I	II	III	
(1)	+	+	+	56
(2)	+	+	−	8
(3)	+	−	+	11
(4)	+	−	−	5
(5)	−	+	+	7
(6)	−	+	−	2
(7)	−	−	+	5
(8)	−	−	−	6
				100

Table 13-6 allows us to answer a great many questions, such as, "How many people did not change their position at all?" The answer comes from groups (1) and (8): 56 percent kept approving throughout (1), 6 percent kept disapproving (8). What was the total approval at each of these interviews? Answer: For interview I add (1), (2), (3), (4), which yields 80 percent; for interview II, add (1), (2), (5), (6), which yields 73 percent; for interview III, add (1), (3), (5), (7), which yields 79 percent (see top line in Table 13-6).

In addition, we can measure the *turnover* from one interview to the other and thus obtain the complete picture, as in Table 13-7.

TABLE 13-7
Summary View of Shifts between Three Interviews

	Interview		
	I	II	III
Total approving	80%	73%	79%
Shift *from approval* at the preceding interview *to disapproval*	. . .	16% (3) + (4)°	10% (2) + (6)
Shift from *disapproval* at the preceding interview *to approval*	. . .	−9% (5) + (6)	16% (3) + (7)
Net shift from preceding interview	. . .	−7%	+6%

° Numbers in parentheses refer to the first column of Table 13-6.

The data allow us also to see in Table 13-8 the distribution of individual shift patterns across the three points in time. The highest approval rate at interview III (88 percent) is found, not unexpectedly, among the 64 percent of the respondents who had approved at both prior interviews I and II; the lowest approval rate (37 percent) belongs to the respondents who had at no prior

TABLE 13-8
Relation between Interviews I, II, and III

At Interview III	Positions at Interviews I and II			
	Approved at I and II %	Approved at II only %	Approved at I only %	Approved neither at I nor II %
Approved	88	78	69	37
Did not approve	12	22	31	63
Total	100%	100%	100%	100%
Share of total sample (100%)	64% (1) + (2)	9% (5) + (6)	16% (3) + (4)	11% (7) + (8)

time approved. The two groups in between, who approved only at one earlier period, provide an additional insight: those who approved at the immediately preceding interview II show a higher propensity to approve at interview III (78 percent) than do those who approved only at interview I (69 percent).

None of these results could have been obtained from consecutive one-shot surveys, which could produce only the marginal percentages of approval of 80, 73, and 79 and, by implication, the difference between these figures, the *net shift*.

OPERATIONAL ANALYSIS

Panel analysis is essential for a thorough understanding of the effects of campaigns, whether they are aimed at voters or at purchasing housewives. Table 13-9 depicts the crucial phases in the marketing campaigns for three competing detergents, at a time when that type of product first came on the market. The figures are derived from a consumer panel in which housewives

TABLE 13-9

Operational Analysis of Three Detergent Campaigns

	Campaign		
	Brand A	Brand B	Brand C
(1) Total campaign costs	$9 million	$6 million	$3 million
Percent of all housewives who:			
(2) tried the product	45%	42%	26%
(3) became regular users	14%	5%	1%
I: (2) ÷ (1) *Advertising effectiveness:* percent of housewives who tried the product per $1 million advertising	5%	7%	9%
II: (3) ÷ (2) *Customer satisfaction:* regular users as percent of those who tried it	31%	12%	4%
III: (3) ÷ (1) *Overall success:* percent of regular users per $1 million	1.6%	0.8%	0.3%

recorded their purchases in diaries that were then collected and analyzed.[3]

The first three lines of Table 13-9 give the basic data for each brand: (1) the amount spent on advertising, (2) the proportion of housewives who tried the product, (3) the proportion who became regular users. The three indices that follow measure the effect of the campaign. (I) Brand C, although it spent less money than the others, reached more housewives (9%) per million dollars advertising than the other two brands. (II) Customer satisfaction was best for brand A; worst for brand C. The bottom line (III), (2) ÷ (3), then reflects the combined effect of promotion and

---

[3] The first permanent consumer panel was established by Sam Barton in the United States. Consumer panels have now proliferated both in the United States and abroad. The first commercial panel operation, a panel of grocery, drug, and five-and-ten-cent stores, became the core of the A. C. Nielsen operation. Strangely enough, in Nielsen's standard reports, these store panels are used primarily as successive independent samples.

product quality: although brand A had the lowest "try" ratio per dollar, it had the highest ultimate success rate, because its conversion rate from trial to regular use was high. In contrast, brand C, the initial persuasion rate of which was high, had the lowest overall success rate because its conversion rate was minimal.

MEASURING LOYALTY

The ultimate goal of all efforts of persuasion is to create loyal customers or loyal voters who need not be persuaded anew when the time comes when they could shift their allegiance. Moreover, as we saw in Table 13-8, loyalty is a property that gains strength with time: the longer it has lasted, the more likely it is to continue. Creating the first loyal event, or preventing the first *dis*loyalty, is the most crucial step in all such efforts of persuasion. The panel allows us to measure loyalty as well as the direction of disloyalty.

Table 13-10 provides an opportunity to study that problem. It is the record of 398,594 automobile purchases, showing the make of the new car as well as that of the car whose place it took.

The table does not represent the full market picture, only that of one-car families in 35 states. It is used here to demonstrate the type of analysis that can be derived from such a switch table even though it gives an incomplete picture of the American car market.

What can we learn from such a table? First, perhaps, by studying the overall balance sheet and comparing the numbers in the bottom row with those in the right-hand column (Table 13-11), we can see that General Motors lost only 1 percent of its market share; Ford, 16 percent; and Chrysler, 10 percent. The great winners were still the imports and, relatively speaking, also American Motors, primarily because their main seller was an Americanized "import," the Renault.

Next we look at the loyalty to a particular make of car. We can compute that measure only for Chevrolet, Ford, and Dodge, because in our table these are the only makes kept separately; the others are grouped in broader categories. From here on we

TABLE 13-10

Brand of Traded-in Car and Brand of Newly Purchased Car

(One-car households; 1984 model year; first quarter; 35 states only)

Traded Old Cars	NEW CARS								
	General Motors		Ford Motor Company		Chrysler Corp.				
	(1)	(2)	(3)	(4)	(5)	(6)	(7)	(8)	(9)
	Chevrolet	Other GM	Ford	Other FMC	Dodge	Other Chrysler	American Motors	Import and Miscellaneous	Total All Old Cars
General Motors									
(1) Chevrolet	27,031	20,079	5,529	2,312	1,857	2,529	1,355	12,327	73,019
(2) Other GM	12,240	70,250	6,658	4,515	2,015	4,132	1,800	17,336	118,946
Ford Motor									
(3) Ford	6,703	11,906	18,158	4,809	1,605	2,271	1,165	9,727	56,344
(4) Other FMC	2,049	5,026	3,265	5,935	445	1,006	421	3,490	21,637
Chrysler Corp.									
(5) Dodge	1,804	2,985	1,445	546	1,819	1,855	388	2,679	14,392
(6) Other Chrysler	2,690	5,271	2,168	969	2,690	5,379	566	3,770	22,632
(7) American Motors	1,033	1,542	953	321	360	471	1,308	1,700	7,688
(8) Import and Miscellaneous	6,480	13,168	5,849	2,260	2,174	2,783	2,368	48,854	83,936
(9) Total of all new cars	60,030	130,227	44,025	21,667	12,965	20,426	9,371	99,371	398,594

Source: *New Car Buyer Analysis* (35 states), R. L. Polk & Co.

TABLE 13-11
Gains and Losses

	Old Car Number	Per- cent Share	New Car Number	Per- cent Share	Per- cent*
Chevrolet	73,019	[18.3]	60,030	[15.1]	−18
Total General Motors	191,965	[48.2]	190,257	[47.7]	−1
Ford	56,344	[14.1]	44,025	[11.1]	−22
Total Ford Motor Co.	77,981	[19.6]	65,692	[16.5]	−16
Dodge	14,392	[3.6]	12,965	[3.2]	−10
Total Chrysler Corp.	37,024	[9.3]	33,391	[8.4]	−10
American Motors	7,688	[1.9]	9,371	[2.4]	+22
Imports and Miscellaneous	83,936	[21.0]	99,883	[25.0]	+19
		100%		100%	

* Old car share = 100 percent.

shall refer to the various cells of Table 13-10 by the two-digit number that is obtained—as is done for areas on a map—by the column and row numbers that are added to the column and row and captions. Thus, the first cell has number 1, 1; the one next to it to the right, 1, 2; the cell in the lower-right corner, number 9, 9. In Table 13-12, then, is the answer to our first brand-loyalty question.

Chevrolet was first; Ford, a close second; and Dodge, a poor third. Next we look at the loyalty of these car owners to their respective corporations. The corporation does not mind customers who switch brands as long as they remain customers of the corporation, especially since such a move often involves shift to a more expensive brand. Table 13-13 gives these figures for the four American carmakers.

These figures, of course, do not only reflect satisfaction with

TABLE 13-12
Loyalty to Leading Brands

	Numbers Buying Same Brand	Number of Previous Owners	Percent Brand Loyalty
Chevrolet	27,031 (1, 1) of	73,019 (1, 9) =	37
Ford	18,158 (3, 3) of	56,344 (3, 9) =	32
Dodge	2,690 (5, 5) of	14,342 (5, 9) =	19

the traded-in car but also the relatively greater or smaller choice of other cars offered by these corporations.

Finally, Table 13-14 allows us to see the balances of trade between the corporations, whereby imports are counted as one such entity.

General Motors had net gains vis-à-vis all its American competitors (I to III) but showed considerable vulnerability to imports (IV). Ford lost to all competitors (I, V to VII), relatively little to Chrysler (V), and much to imports (VII). Chrysler lost to GM (II), American Motors (VIII), and imports (IX), but had a slight gain over Ford. American Motors lost only to GM (III) and gained from the three others (VI, VII, IX). Imports lost only to American (X) and gained from all others (IV, VII, IX).

The intriguing question behind all such shift tables is—"Why?" Why a shift at all? Why to that particular make? I do not have these answers; one hopes the carmakers have them. The following section discusses some of the general roads to such answers.

TABLE 13-13
*Corporation Loyalty: Percent Buying Cars
Made by the Same Corporation*

General Motors	68
Ford Motor Company	41
Chrysler Corporation	32
American Motors	17

TABLE 13-14

Trade Balances between Corporations

(Read the % figures as follows, e.g., for (I): Of the 43,698 cars that shifted between Ford and GM, 44% shifted from GM to Ford, and 56% from Ford to GM, a net gain of 12% for GM.)

(I) GM and Ford traded 43,698 cars		(II) GM and Chrysler traded 23,303 cars		(III) GM and American Motors traded 4,730 cars	
Ford	44%	Chrysler	45%	American Motors	45%
GM	56%	GM	55%	GM	55%
GM	+12%	GM	+10%	GM	+10%

(IV) GM and imports traded 49,311 cars		(V) Ford and Chrysler traded 10,455 cars		(VI) Ford and American Motors traded 2,860 cars	
Imports	60%	Chrysler	51%	American Motors	55%
GM	40%	Ford	49%	Ford	45%
GM	−20%	Ford	−2%	Ford	−10%

(VII) Ford and imports traded 21,326 cars		(VIII) Chrysler and American Motors traded 1,785 cars		(IX) Chrysler and imports traded 11,406 cars	
Imports	62%	American Motors	53%	Imports	57%
Ford	48%	Chrysler	47%	Chrysler	43%
Ford	−24%	Chrysler	−6%	Chrysler	−14%

(X) American Motors and imports traded 4,068 cars

Imports	42%
American Motors	58%
American Motors	+16%

WHO SHIFTED AND WHY

In a way, every controlled experiment is a panel operation, since it requires observation of the same subjects before and after the experiment. In nonexperimental situations, the panel offers the best opportunity to explore causes and effects, because it establishes the precise sequence of events over time and thus allows us to relate shifts to prior exposures and other influences and permits precise focusing on individual behavior.

A panel cross tabulation is not different from any other cross tabulation. But because the panel as a rule provides more information on the preceding causal chain than does the one-shot survey, it can better guard against spurious correlations and can provide insights superior to those derived from normal survey data.

Table 13-15 provides more data on the Willkie-Roosevelt presidential campaign of 1940.[4] It relates the vote intention expressed in August to the vote intention in May and to the campaign influences during the intervening months. We begin with line (1) of Table 13-15: In May, 55 percent of the panel members intended to vote Republican and 45 percent, Democrats. Line (2) records the proportion of panel members who exposed themselves primarily to either the campaign of the Democrats or that of the Republicans, or to neither or both: 25 percent (10 and 15) of the panel members were reached predominantly by the Democratic campaign; almost twice as many $(32 + 17) = 49$ percent were reached by the Republican campaign.

Line (3) shows that even the Republicans who were exposed primarily to the Democratic campaign showed only minor defections—7, 3, and 3 percent, respectively. Defections among the May Democrats were more frequent: 7, 8, and 20 percent, respectively. Note that the shift for each group is largest among

[4] P. F. Lazarsfeld, B. Berelson, H. Gaudet, *The People's Choice* (New York: Columbia University Press, 1948).

TABLE 13-15
Exposure to Presidential Campaign and Shift in Vote Intention

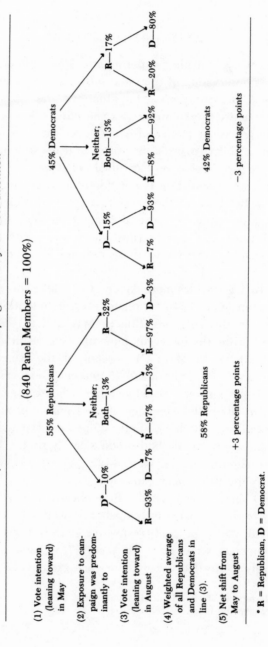

(840 Panel Members = 100%)

(1) Vote intention (leaning toward) in May

55% Republicans   45% Democrats

(2) Exposure to campaign was predominantly to

D*—10%   Neither; Both—13%   R—32%   D—15%   Neither; Both—13%   R—17%

(3) Vote intention (leaning toward) in August

R—93% D—7%   R—97% D—3%   R—97% D—3%   R—7% D—93%   R—8% D—92%   R—20% D—80%

(4) Weighted average of all Republicans and Democrats in line (3).

58% Republicans   42% Democrats

(5) Net shift from May to August

+3 percentage points   −3 percentage points

* **R** = Republican, **D** = Democrat.

those who were reached primarily by the campaign of their May opponents—7 percent as opposed to 3 percent—among the Republicans, but 20 percent as compared to 8 percent among the Democrats if they were exposed to the Republican campaign. Without more data, one cannot conclude that it was a superior Republican campaign that produced the net shift, since it is quite possible that the shift represents those panel members whose May intention had been less certain and therefore chose to have greater exposure to their opponents' campaign.

Compared with successive one-shot surveys, such panel data on exposure and shift of individual subjects provide superior conditions for interviewing for reasons. How did the exposures come about? To what degree were they self-selected? Did these exposures contribute to the shift?

THE BANDWAGON EFFECT

Voters who had not yet made up their mind during the May interview were asked which party they expected to win the presidential election. By October, most of these people had made up their mind. Table 13-16 shows how the expectation about who would win was related to their eventual voting intention.

TABLE 13-16
Vote Intention and Expectation of Winner

Vote Intention in October	No Vote Intention in May but Expected Republicans to Win %	No Vote Intention in May but Expected Democrats to Win %
Democrat	43	69
Republican	52	31
Total	100	100

Some people tend to vote for the candidate they expect to win, a fact commonly known as the bandwagon effect. It is corroborated by what the respondents themselves had to say about why they switched; for example, "I have always been a Democrat, but lately I've heard of so many Democrats who are going to vote Republican that I might do the same. Four out of five Democrats I know are doing that."

EFFECTS OF ADVERTISING

Before showing how the panel can help in focusing on cause-and-effect relations, it will be useful to look at the type of simple-minded evidence that is frequently presented as proof of a causal relation. Table 13-17 relates remembering exposure to an advertisement to the use of the advertised product.

TABLE 13-17
Advertising and Use of Product

	Remember Having Seen the Advertisement		Don't Remember Having Seen the Advertisement	
	Number	Percent	Number	Percent
Use advertised product	(188)	30	(298)	21
Don't use advertised product	(434)	70	(1,150)	79
Total	(622)	100	(1,448)	100

Table 13-17 seems to give great credit to the effectiveness of the advertisement: the proportion of users is almost 50 percent higher among those who remember being exposed to the advertisement (30 percent) than among those who do not recall such an exposure (21 percent). The mistake of Table 13-17 lies in its failure to distinguish use of the advertised product *before* and *after* exposure to the advertisement, as in Table 13-18.

TABLE 13-18

Product Use Before and After Exposure to Advertising

Used Product after	Used Product Before		Did Not Use Product Before	
	Saw Advertise- ment %	Did not see Advertise- ment %	Saw Advertise- ment %	Did not see Advertise- ment %
Yes	72	71	10	10
No	28	29	90	90
Total	100	100	100	100
Number	(198)	(252)	(424)	(1,196)

It turns out that the advertisement actually left no traceable effect. The spurious correlation in Table 13-17 was obtained because the people who were users of the advertised product were more likely to notice the advertisements (198 out of 198 + 252 = 44 percent) than those who had not been users (424 out of 424 + 1,196 = 26 percent). The crucial element neglected in Table 13-17 but introduced in Table 13-18 is the time factor.

The example above is not contrived but is in fact the prototype of an error frequently seen. The size of advertising budgets, with little regard for considerations of effectiveness, is traditionally determined by the current profit situation. When profits are low, the budget shrinks; when profits are high, budgets expand and thereby create a misleading link between size of the advertising budget and profits.

The panel helps one avoid mixing up temporal sequences, as shown by the following analysis. A cross section of television-set owners was interviewed in February and again 3 months later in May of the same year.[5] Each time it was ascertained whether or

[5] From an unpublished study by Hugh M. Beville, Jr., *Why Sales Come in Curves.* Based on a study made by the old Bureau of Applied Social Research for the National Broadcasting Company.

TABLE 13-19
Viewing and Buying Pattern Combining February and May

	VIEWING			BUYING	
	In February			In February	
In May	Yes	No	In May	Yes	No
Yes	+ + Continued	+ − Stopped	Yes	+ + Continued	+ − Started
No	− + Started	− − Never listened	No	− + Stopped	− − Never bought

not the respondent had been viewing a certain program and whether she had been buying the product advertised on it.

The ++ are individuals who continued; the +−, ones who stopped in the second period; the −+, ones who started in the second period; and the −−, ones who never started viewing or buying. Each individual would then have to fall into one of the four viewing patterns and one of the four buying patterns; by combining the two we obtain (4 × 4) = 16 possible viewing and buying patterns. Table 13-20 is this 16-cell table.

The individuals (82) in the first cell viewed and bought in both periods (++, ++); the second cell downward contains the individuals (53) who continued viewing (++) but stopped buying (+−); and so forth. The four columns of numbers are then also translated into percentages, considering viewing as the cause of buying.[6]

We compare first the persons who *started viewing* the program after February (−+) with those who *never viewed* it (−−). In this way we can assess the effect of *having started to view*.

1.   Gaining new buyers: Among those who started viewing, 8.4

---

[6] See Chapter 3.

TABLE 13-20

Viewing and Buying in February and May*

Buying	VIEWING								Total Buying
	Continued (++)		Stopped (+−)		Started (−+)		Never Viewed (−−)		
	Number	Per-cent	Number	Per-cent	Number	Per-cent	Number	Per-cent	
Continued (++)	(82)	12.0	(31)	9.8	(27)	9.2	(104)	9.0	(244)
Stopped (+−)	(53)	7.7	(28)	9.0	(20)	6.8	(80)	6.9	(181)
Started (−+)	(57)	8.3	(24)	7.6	(24)	8.4	(81)	7.0	(186)
Never bought (−−)	(481)	72.0	(231)	73.6	(219)	75.6	(891)	77.1	(1,822)
Total viewing	(673)	100.0	(314)	100.0	(290)	100.0	(1,156)	100.0	(2,433)

* The first + or − sign refers to February; the second sign to May.

percent started to buy as compared with 7.0 percent among those who never viewed—an increase of 1.4 percentage points, or (1.4 over 7.0) = 20.0 percent.

2. Holding old buyers: Among those who started viewing, 6.8 percent stopped buying as compared with 6.9 percent among those who never viewed—an increase of 0.1 percentage points, or (0.1 over 6.9) = 1.4 percent.

By combining effects 1 and 2, we obtain the overall gain attributable to *having started to view* the program:

3. Among those who started viewing, the percentage of buyers increased from February to May from 16.0 percent (9.2 plus 6.8) to 17.6 percent (9.2 plus 8.4), that is, by 1.6 percentage points, or 10.0 percent (1.6 over 16.0).

4. Among those who never viewed, the percentage of buyers increased from 15.9 (9.0 plus 6.9) to 16.0 percent (9.0 plus 7.0), that is, by 0.1 percentage points, or 0.6 percent (0.1 over 15.9).

5. Hence, having started to view increased the number of buyers in that group by 9.4 percent (10.0 minus 0.6) over what it would have been had they not started.

Similarly, by comparing those who continued viewing with those who stopped viewing, we assess the effect of having continued viewing.

6. Gaining new buyers: Among those who continued viewing, 8.3 percent started to buy compared with 7.6 percent among those who stopped viewing—an increase of 0.7 percentage points, or 9.2 percent (0.7 over 7.6).

7. Holding old buyers: Among those who continued viewing, 7.7 percent stopped buying compared with 9.0 percent among those who stopped viewing—a decrease of 1.3 percentage points, or 14.4 percent (1.3 over 9.0).

By combining effects 6 and 7, we can determine the overall gain attributable to *having continued viewing* the program:

8. Among those who continued viewing, the percentage of buyers increased from February to May from 19.7 (12.0 plus 7.7) to 20.3 (12.0 plus 8.3), that is, by 0.6 percentage points, or 3.0 percent (0.6 over 19.7).
9. Among those who stopped viewing, the percentage of buyers decreased from 18.8 (9.8 plus 9.0) to 17.4 (9.8 plus 7.6), a decline of 1.4 percentage points, or 7.4 percent (1.4 over 18.8).
10. Hence, continued viewing increased the number of buyers in that group by 10.4 percent (3.0 plus 7.4) over what it would have been had they stopped viewing.

We can now move to the final step, the overall evaluation of the television program in terms of the added number of buyers of the advertised products as compared with what that number would have been without the programs. Here we must consider not only the *effect* of started and continued viewing, as compared with stopped and never viewing, but also the *frequency* with which people started and continued viewing. We return, therefore, to Table 13-19 with this consideration: had all the 290 people who actually started viewing stopped (e.g., if the program had been canceled), they would have reacted as did the 1,156 people who never viewed. The number of May buyers among the 290 was 51 (27 + 24); had this group of 290 not started viewing, there would have been four fewer buyers according to statement 5 above.

Similarly, we can compute the loss of buyers among the 682 people who continued viewing for the hypothetical case that they had stopped viewing as the 314 people actually did. Had the 139 May buyers (82 + 57) not continued viewing, they would have been reduced by 13, according to statement 10.

We now compare the three groups of figures in Table 13-21.

TABLE 13-21

The Overall Effect

Actual Buyers in February 425	Actual Buyers in May 430	Computed Buyers in May (If Program Had Not Been Available) 413
The totals of lines 1 and 2 in Table 13-19 (244 plus 181). Expressed as percentage of the total sample (2433) this is	The totals of lines 1 and 3 (244 plus 186). Expressed as percentage of the total sample this is	430 minus the 17 (4 plus 13) buyers who would have been lost had the program been canceled. Expressed as a percentage of the total sample this is
17.5%	**17.7%**	**16.9%**

The overall effect of having had the program continued through May was thus an increase among buyers from the hypothetical 16.9 to the actual 17.7, an increase of 0.8 percentage points, or 4.7 percent, over the February level.

REVERSAL OF CAUSE AND EFFECT

Our 16-fold table relating viewing and buying at two different time periods also permits answering a collateral question which has often puzzled researchers: is there possibly not only an effect of viewing on buying (as we have so far assumed) but also a reverse effect of buying on viewing; may not buyers of a particular brand tend to view the sponsor's program more frequently than do nonbuyers? A *feedback* factor could thus contribute to the correlation found between buying and viewing.

By comparing February buyers and nonbuyers, the panel data offer two tests of the feedback hypothesis.

1. February buyers more than nonbuyers should claim that they

started viewing between February and May.
2.   February buyers more than nonbuyers should claim that they
     continued viewing the program after February.

A significant difference in the proportion of buyers and nonbuyers who report starting and continuing to view would suggest
that there was indeed a reversal of cause and effect: we first test
whether there are more claims of *starting* to view among buyers
than there are among nonbuyers. We divide the 1,446 nonviewers
in February into two groups: those who were buyers in February
and those who were not. We then compute in Table 13-21 for
each group the proportion who report in May that they had
started to view between February and May.

If it were true that buyers were more apt to start viewing than
were nonbuyers, a greater proportion of buyers should report
starting to view during the 3-month period (test I). No such
difference is apparent.

Test II examines whether proportionately more buyers than
nonbuyers continue viewing. Here we take the 996 persons who
were viewers in February and trace their later viewing separately
for buyers and nonbuyers.

Here too the difference in viewing is insignificant. If feedback
were operating, we should find a higher proportion of continued
viewing among the February buyers.

Since neither of the two tests shows a significant difference, we
conclude that there was no feedback effect that caused buyers to
view. A feedback is conceivably more likely after the purchase of
a more substantial product, such as an automobile.

BIAS FROM PRIOR INTERVIEWING

The very fact of having been interviewed before about a given
topic may make the respondent self-conscious and influence the
result of the subsequent interviews. This effect is different from

TABLE 13-22

Testing the Feedback Hypothesis

	February Buyers	February Nonbuyers
I		
(a) Number who were nonviewers in February	231	1,215
(b) Number of February nonviewers who started to view between February and May	47	243
(c) As percentage of (a)	20.3%	20.0%
II		
(a) Number of viewers	19	793
(b) Number of February viewers who continued viewing after February	135	538
(c) As percentage of (a)	69.6%	67.8%

the "interviewer effect," the continuing contact with a particular interviewer.

That this danger exists is obvious. But how great a danger is it? Generally speaking, the danger will be small if the topic is one which the respondent has been well aware of anyway; if it is a topic she had not thought about before, the danger may be considerable.

The panel offers the opportunity for an experimental design that will disclose whether or not such bias operates. The main panel is reinterviewed while the reinterview in a corresponding control panel is omitted; or the main panel is interviewed periodically several times while the control panel is interviewed only twice, once at the time of the first interview and again at the time of the last interview. Whatever differences are then observed between the two groups beyond the normal sampling error can be attributed to the interviewing effect.

Table 13-23 reports on a series of such experiments in which parallel questions were asked of a panel and in successive one-shot surveys of approximately twice the size of the panel. The main panel consisted of 425 respondents; three of the reported questions (1, 4, and 9) were based on a larger panel of 728 respondents.[7]

On the whole, the changes in the panel and in the control survey run in the same direction and are of similar magnitude; the rank-order coefficient of correlation is .77.

Three questions—3, 5, and 13—yield somewhat larger discrepancies. Two of them are of the kind that are not likely to have occurred to the respondent prior to the interview; hence, on our theory, they might be expected to bias the reinterview.

Experiments made to test the reliability of a panel used in the 1940 presidential campaign confirmed the hypothesis that no significant differences between panel and control group are to be expected if the particular question is part of the general public discussion. No panel bias could be discovered on the following questions:

For which party's candidate do you think you will probably vote this fall?

Which party do you think will actually be elected?

What kind of president do you think Willkie would make: good, fair, or no good?

There is some evidence that while prior interviewing does not affect the *direction* of a change, it may affect its size. In Table 13-24 the changes between the two independent samples were found to be somewhat greater than were those within the panel. One might speak of a "freezing" effect of the panel. If the

[7] The study was conducted during World War II, between December 1941 and June 1942, by the Bureau of Applied Social Research at Columbia University in cooperation with the Gallup Poll on a grant from the Rockefeller Foundation.

TABLE 13-23

*Panel Changes as Compared with Corresponding Changes in
Independent Field Samples*

Question	Inter-viewing Interval	Percentage Refers to Those Saying:	Percent Change*	
			In Panel	In Control Sample
1. "If Hitler offered peace now to all countries on the basis of not going further, but of leaving matters as they are now, would you favor or oppose such a peace?"	December to June	"Favor"	+5.6(1)	+5.0(3)
2. "So you think the U.S. is doing all it possibly can to win the war?"	February to June	"Yes"	+5.2(2)	+5.5(2)
3. "Whom do you consider to be the chief enemy: the German government or the German people, or both?"	December to June	"Both"	+4.6(3)	+1.8(6)
4. "If Hitler offered peace now, to all countries on the basis of not going any further but on leaving matters as they are now, would you favor or oppose such a peace?"	December to February	"Favor"	+4.3(4)	+2.2(5)
5. "Are you satisfied with the conduct of the war against Japan?"	February to June	"Yes"	+3.4(5)	+8.0(1)
6. "If Hitler offered peace now to all countries on the basis of not going further but of leaving matters as they are now, would you favor or oppose such a peace?"	February to June	"Favor"	+2.2(6)	+3.6(4)

TABLE 13-23 *(Continued)*
Panel Changes as Compared with Corresponding Changes in
Independent Field Samples

Question	Inter- viewing Interval	Percentage Refers to Those Saying:	Percent Change*	
			In Panel	In Control Sample
7. "Whom do you consider the chief enemy: the German government or the German people?"	February to June	"German people"	+1.2(7)	+0.9(7)
8. "Do you approve of the President's home policy?"	December to June	"Yes"	+0.9(8)	−2.8(11)
9. "Are you satisfied with the conduct of the war against Japan?"	December to June	"Yes"	−0.2(9)	−0.7(9)
10. "Do you approve of the President's home policy?"	February to June	"Yes"	−2.8(10)	+3.8(12)
11. "Do you approve of the President's home policy?"	December to June	"Yes"	−4.0(11)	−7.0(10)
12. "Are you satisfied with the conduct of the war against Japan?"	December to February	"Yes"	−5.3(12)	−10.5(13)
13. "Whom do you consider to be the chief enemy: the German government or the German people?"	February to June	"German govern- ment"	−6.7(13)	+0.4(8)

* The figures in parentheses indicate rank of change, in order of size and direction. Greatest positive change has rank 1; greatest negative change has highest rank 13. There are altogether 13 questions.

changes in the panel are greater than those in the control group, we might speak of a "stimulating" effect. Something like it could be observed in the repeatedly cited presidential campaign panel of 1940. There, each member was asked on six different occasions for whom he intended to vote. Table 13-24 shows one change between the first and sixth interview in the panel, compared with the control group that was interviewed only twice: once at the time of the first panel interview and again at the time of the sixth interview.

TABLE 13-24

Speeding up Effect of Repeated Interviewing

(Those who had not yet decided at time of first interview)

	Had Been Previously Interviewed Five Times	Had Been Previously Interviewed Only Once
Know for whom they will vote	60%	45%
(Number of respondents = 100%)	(213)	(214)

Repeated interviews seemed to have the effect of speeding up the process of deciding for whom to vote.

A somewhat more complicated experiment was designed to measure the effect of a preceding interview on three items: on the self-selection of a subsequent exposure (viewing a television program), on the evolving crucial attitude, and on the effect of that exposure on the attitude. Figure 13-1 depicts this design.

By comparing (a) with (a'), (b) with (b'), and (c) with (c'), one can determine the influence of prior interviewing on the self-selected exposure. By comparing the relation of (a, b, c) to (A') with the relation of (a', b', c') to (B'), one can determine the influence of prior interviewing on the effect of the exposure on subsequent attitude; and by comparing (A') with (B'), one has an overall measure of the combined effect of these two possible influences.

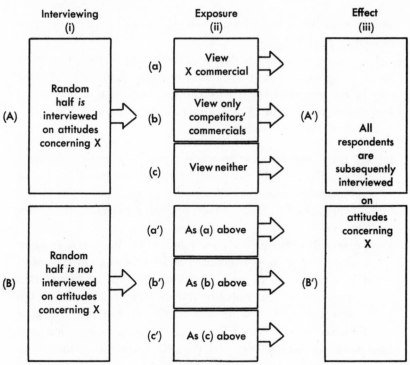

FIGURE 13-1
*Measuring Interviewing Bias on Exposure and on the
Effect of That Exposure*

PANEL MORTALITY

Almost any effort to sample a human universe is bound to result
in some failures. Only nonliving matter allows perfect sampling.
Humans may be sick, temporarily absent, or inaccessible, or they
may simply refuse to cooperate. All these problems are aggravated
in a panel operation. If the respondent is told that he will be
interviewed more than once or that he is required to keep records,
the refusal rate will increase. At subsequent interviews, the losses
that affected the first round of interviews—temporary absence
and refusal to cooperate—are bound to recur and to accumulate.
In addition, there will be losses specific to a panel, resulting

simply from the elapsed time interval: respondents may have moved or died.[8]

There is no need here to deal with the bias problems of ordinary survey samples; they have been discussed elsewhere with sufficient care.[9] Our concern here is with the specific bias that may result from panel operations. A few generalizations garnered from experience will have to suffice:

Losses from mobility occur more frequently among younger people than among older ones, and more frequently in larger communities than in smaller ones.

Losses from actual mortality affect, of course, primarily the oldest age brackets.

Temporary absence affects men more often than women, and more often the upper income brackets than the lower ones.

Refusals to cooperate depend greatly on the type of cooperation required. If it is onerous, the normal refusal rate will be high unless a reward of some sort is offered. But the reward is then likely to be more effective in the lower income brackets.

Mail panels requiring literacy skills, however modest, suffer relatively greater losses in the lowest educational and income brackets. This bias does not necessarily evolve out of an initial refusal to cooperate; it may result from the subsequent failure to provide the promised cooperation.

SUMMARY

Obtaining continual information from response units over time offers a variety of advantages: it increases the sheer amount of information; it increases the accuracy of the information by reducing the memory burden; it also permits a more precise

[8] To guard against loss through unreported moving, one should secure at the first interview names and addresses of relatives or friends, who later on might help to find the lost panel member.

[9] See "Nonsampling Errors," *The Encyclopedia of the Social Sciences* (New York: Macmillan, 1968).

determination of processes, such as use habits that are derived from patterns that extend over time. Most importantly, it makes possible a sophisticated analysis of shifts and changes and a tying up of these changes with prior events that have brought them about. Like all such tools, panel technique has its dangers— conditioning of the response units, and both actual and administrative mortality. But both these dangers can be guarded against. The panel has thus become one of the more powerful tools of causal analysis.

14

Triangulation

Most efforts of scientific proof suffer from one shortcoming or another. Because of physical, social, and conceptual limitations, our research efforts are forever imperfect. This is true for all sciences but especially the social sciences, where even controlled experiments are no exception. Not only do experiments have margins of error, they are conducted at one place, at one point in time, with one target population. Thus questions are raised about the breadth with which their results can be generalized. Some of these uncertainties are reduced by a frequently encountered constellation of proof—the confluence of evidence from two or more independent research approaches, a constellation for which the term *triangulation* has been receiving some acceptance.

The term originated in the surveyor practice of determining the location of terrain points through successive delineation of triangles. It was then expanded to cover the method navigators use to determine their ship's location.

We would take a star sighting and from that you get a line you could draw on the chart that represented a line on the face of the Earth. That one line wasn't very useful to us. But if we could find a line from another star and cross it and then another one and it rarely crossed at the same point, you began to get a succession of lines that would define generally the area which we might expect the ship to be at a particular point in time.

More recently the term has come to designate any scientific effort to approach the truth of a proposition through more than one independently developed research channel.

Philosopher Herbert Feigl was perhaps the first person to use the term in its transformed sense. Donald Campbell and Donald Fiske later provided the first systematic application of the triangulation method and thus firmly established its expanded meaning.

The more formal approach developed from a practice that is common to research efforts in all sciences, especially if they are concerned with objects that are distant in space or time. Efforts have been made to reconstruct the true state of affairs from bits and pieces which, if properly put together, suggest the resolution of the puzzle.

What follows here is an attempt to describe some of the triangulation opportunities with two purposes in mind: to make visible the variety of these opportunities and to stimulate thereby the invention of new variants of this approach.

REPEATED OBSERVATIONS

The established routine of repeating a measurement or other observations in order to increase their accuracy is a form of triangulation. The average of repeated measurements has a greater claim to accuracy than its individual components.

A variant of this principle is employed when an operation relies on personal judgments. Coding, that is, the classification of obtained answers to a survey question, is an example. Personal judgments may vary here in marginal situations. The way to increase the objectivity of coding is to have each answer coded twice independently by two persons. Points of disagreement are then highlighted and resolved either by consent or a decision by a supervisor.

GAUGING IMPERFECT SAMPLES

Surveys and polls normally aim at a probability sample base that permits projection of their findings to the sampled universe. Such sampling efforts are often marred by imperfections the impact of which is difficult to gauge. At times the task is made easier if some of the survey findings can be collated with their counterpart established by a different source, the reliability of which is not questioned.

A survey of pharmacists in the United Kingdom was conducted with the aim of eliciting their experiences with different brands of a certain product. When questions arose concerning the impact of some sampling deficiencies, the following constellation proved helpful. As a starting question, the pharmacists were asked which of the brands they carried and how many units of each they sold in an average month. The data allowed the computation of two sets of numbers, both of which were available through the bimonthly sampling audits of pharmacies by the A. C. Nielsen Company: the market share of the various brands and the percentage of pharmacies that carried each brand. When the survey figures turned out to reasonably match the corresponding Nielsen data, confidence in the unbiased nature of the survey sample was strengthened.

The inclusion of questions that allow controlling comparison with reliable outside sources is a recommended safety device in survey making.

TRIANGULATING SIMULATED RESULTS

Experimental designs must at times use simulations of one sort or another,[1] which always raises doubts about whether such experiments can be accepted as representing the real event. At times, the question can be answered with some precision.

A problem, called sentencing disparity, arises in criminal law proceedings. Judges have broad discretion to determine sentences; and some judges are more severe sentencers than are others. Actual figures on such disparities are hard to come by, since each convicted offender is being sentenced by only one judge and we have no way of knowing with precision how he would have fared before another judge. When judges in the United States Court of

[1] See Chapter 8.

Appeals for the Second Circuit decided to study the problem, I suggested that the files of a number of convicted offenders, with all the information available to the sentencing judge, be circulated to *all* judges in the court, asking them what sentence they would have imposed on the defendant. The Federal Judicial Center then carried out this experiment, which was designed to measure sentencing disparity, always with the lingering doubt of whether such simulated sentencing reflected reality.

Some years later, the opportunity arose of obtaining a measure of sentencing disparity in the federal courts through the activity of so-called sentencing councils. Some federal courts have councils in which the trial judge is routinely joined by two of his colleagues who tell him in an advisory capacity what sentences they would impose in a particular case. The judges engage in real sentencing decisions, although only in a preliminary fashion. From both the Second Circuit study and from the sentencing council data, the same disparity measure can be obtained, namely, the average (mean) difference in the sentence of the same defendant by two judges, selected randomly from a particular court. In the sentencing experiment the difference was +48 percent (taking the lower sentence as 100 percent); in the sentencing council data, the difference was 46 percent, which is evidence that showed that the simulated sentencing study did report reality.[2]

EXPERIENCES IN DIFFERENT PLACES

Sometimes institutions may develop in places far apart and under somewhat different circumstances. They occasionally yield data on their functioning that allow reasonable inferences about how that institution will fare in a new location.

When the British Parliament considered abandoning the una-

---

[2] Shari S. Diamond and Hans Zeisel, "Sentencing Councils: A Study of Sentence Disparity and Its Reduction," *University of Chicago Law Review*, vol. 43, 1975, pp. 109, 146. On the general problem see Marvin E. Franbol, *Criminal Sentences; Law Without Order* (New York: Hill and Wang, 1973).

TABLE 14-1
Last Vote at Which Juries Were Hung

Not Guilty:Guilty	Percent
11:1	24
10:2	10
9:3	
3:9	58
2:10	8
1:11	. . .
	100

nimity rule for jury verdicts and allowing 11-to-1 and 10-to-2 verdicts, Harry Kalven and I were asked by how much that change would reduce the frequency of hung juries.[3]

We made the estimate as a result of looking at three independent sets of data. None alone would sustain a prediction, but together they did. We first looked at the frequency of hung juries in the state of Oregon, which had the 10-out-of-12 rule which England was about to introduce, and compared it with the hung jury frequencies in states that required unanimous verdicts. Oregon had 3.1 hung juries for every 100 trials, as opposed to 5.6 hung juries in the states that required unanimity. The figures suggested that the change should reduce the frequency of hung juries from 5.6 to 3.1, that is, by 2.5 percent. Expressed as a fraction of all hung juries, the change would reduce the number by some (2.5 of 3.1) = 45 percent.

A second set of data showed for a sample of hung juries in courts that required unanimity the number of jurors who prevented a verdict by refusing to join the majority.

Table 14-1 shows that the 10-to-2 and 11-to-1 hung juries represent (24 + 10 + 8) = 42 percent of all hung juries. Since

---

[3] We had discussed this problem in *The American Jury*.

11 and 10 are the majorities which under the new rule would be allowed to render a verdict, the 45 percent reduction, calculated from a completely different set of data, strikingly corroborated the Oregon finding.

There was yet a third set of data that had bearing on this issue. Some years earlier, the state of New York abolished the unanimity rule for jury trials, albeit in civil cases, by allowing 11-to-1 and 10-to-2 majority verdicts. During the year before the change, there were 5.2 hung juries per 100 trials; during the 6 years after the rule was changed, the number shrank to 3.1 per 100 trials. The reduction amounted to $(5.2 - 3.6 \text{ of } 5.2) = 31$ percent—a third, independent source for the estimate.

The three independent sources yielded the very similar estimates of 45, 45, and 31 as the percentage by which the reform would reduce the number of hung juries and thereby increased our confidence in the correctness of the prediction.

CROSS EXAMINATION

If a survey is suspect, cross examination, as is done in a court of law, might destroy or, as the case may be, confirm it. Even in a nonhostile environment, triangulation efforts will prove helpful. In a study that explored reasons for moving, some families gave as their primary reason lack of sufficient space in the old lodgings. These people were later asked whether the number of household members had increased prior to their moving.

The reason given for moving is buttressed by the reported growth of the household. That the association is not perfect should not be a surprise, since the feeling of being crowded might also be a result of other circumstances.

A similar situation arose when women gave as a reason for not buying in a certain department store that it was a "poor store." In a later part of the interview, the women were asked to evaluate the store on the basis of a variety of characteristics. The number of specific complaints was then related to the frequency with

TABLE 14-2
Verification of Reasons for Moving*

	Among Those Who Had	
	Prior Increase in Size of Household %	No Change in Size of Household %
Percent giving space limitations as main reason for moving	80	30

* From Peter Rossi, *Why People Move* (New York: Free Press, 1955), p. 144.

which women had given "poor store" as a reason for not buying. Table 14-3 shows the result.

There is slight puzzlement about women who speak of a "poor store" without reporting a specific complaint. They probably base their judgment on hearsay without having an experience of their own. An interviewer might well try finding out whether or not they indeed ever went shopping in that store.

TABLE 14-3
Verification of Reasons for Not Buying in Store X*

Number of Specific Complaints about the Store	Percent of Women in Each Group Giving "Poor Store" as a Reason for Not Buying There
None	25
One	48
Two	64
Three or more	78

* From Paul F. Lazarsfeld, "Evaluating the Effectiveness of Advertising by Direct Interviews," in Paul Lazarsfeld and M. Rosenberg (eds.), *The Language of Social Research* (New York: Free Press, 1955), p. 411.

FAILURE TO AGGREGATE THE EVIDENCE

After Sacco and Vanzetti were convicted of murder and robbery and sentenced to death, their counsel petitioned the governor of Massachusetts to commute the death sentence. When important voices began raising doubts about the propriety of the trial proceedings, Governor Fuller appointed a commission to advise him. The issue before the commission was not the usual one, namely, to determine whether there were sufficient mitigating circumstances to warrant commutation, but the rare and more important one, whether there was sufficient doubt—the British formulation of the measure is a "scintilla of doubt"—of whether or not the convicted defendants actually committed the crime. The report of the commission, chaired by the then president of Harvard, advised against commutation. It had reviewed the six procedural defects alleged to have marred the trial and discussing each in turn concluded that none was sufficient to warrant commutation. The report came as a surprise and puzzle to many. Some time later, the philosopher John Dewey tried to resolve the puzzle. In an essay on "Psychology and Justice"[4] he pointed to the logical fallacy to which the commission succumbed. It considered and treated the merits of each of the six claimed defects separately and independently from each other instead of perceiving them as different symptoms of a common cause—prejudice against defendants who were anarchists. Dewey argued that had the commission considered the joint weight of the six defects instead of weighing each separately, it would have been forced to arrive at a different conclusion.

CORRECTING FALSE COUNTS

Triangulation also helps to resolve less weighty questions. When it comes to reporting the age of very old people, both they themselves and their family members tend toward exaggeration,

[4] John Dewey, *Characters and Events* (New York: Octagon, 1970), vol. II, p. 526.

which poses a problem for the Bureau of the Census, which would like to know with precision the number of persons in their 100th year, 101st year, and so forth. To correct for this inaccurate reporting, the count must be compared with the results from the following independent sources: (1) The number of persons reported as 90 years and older at the previous census (10 years earlier) minus the number of reported deaths in that age group; (2) the standard survival expectations applied to the subgroups of that 90-and-over age group; (3) finally, Medicare records. From the confluence of these sources, the best estimate was developed of the number of people who survived their 100th birthday.[5]

A different accuracy problem arises when we must improve numerical estimates that differ depending on who makes the estimate, which is often admittedly not more than an educated guess. One of those not unimportant numbers is the value of the goods stolen each year by drug addicts. For New York City, the amount had been estimated to be in the $1 billion range. An effort has been made to arrive at a supportable estimate through triangulation. One approach has been to estimate what may be called the demand: the number of addicts and the cost to them of buying heroin. This calculation has been made for two markets: the lower price for the insider (the addict who is also a seller) and that for the outsider (who must pay the higher market price). The total value of stolen goods has next been estimated both from crime statistics and from the losses through shoplifting which are not reported to the police but are estimated by the merchants. Each step has required a number of subestimates, which have been, in turn, refined by triangulation. When it is all done, the heroin-caused thefts turn out to be closer to $250,000 rather than the billion dollars they were thought to be.[6]

---

[5] See also Chapter 3.

[6] Max Singer, "The Vitality of Mythical Numbers," *The Public Interest*, vol. 9, 1971, p. 3.

THE CONVICTION-PRONE JURORS

Not long ago, two federal district courts acknowledged the principle of proof through triangulation. The issue had to do with whether or not a jury from which all absolute opponents of the death penalty were excluded in a capital trial was the "impartial jury" mandated by the Sixth Amendment to the Constitution. Proof that it was not was submitted by way of several independently conducted studies that showed that jurors who favored the death penalty had a higher propensity to decide the issue of guilt in favor of the prosecution than did jurors who were against the death penalty. Since our laws could not possibly allow trying a criminal case by way of a controlled randomized experiment, all submitted studies had imperfections of one sort or another. The first of these studies, conducted in 1953 and published in 1968, involved real jurors and real verdicts in the courts. But this was not a controlled experiment; the controls were introduced only into the analysis.[7] Subsequent studies of the same problem, by different authors were controlled experiments. But these studies have another imperfection: they simulated jury trials with simulated jurors. The last in this series of controlled experiments, however, closely approached the real situation: persons eligible for jury duty were shown a videotaped enactment of a real murder trial, with the evidence in balance so as to make acquittal and conviction equally plausible. In this experiment again, as in all the others, the jurors who favored the death penalty were more likely to convict.[8]

This series of altogether eight studies, conducted independently by a variety of authors, was submitted as evidence that in the guilt phase of a capital trial the exclusion of absolute opponents

[7] H. Zeisel, *Some Data on Juror Attitudes Toward Capital Punishment*, Monograph, Center for Studies in Criminal Justice, University of Chicago 1968.

[8] C. Cowan, W. Thompson, P. C. Ellsworth, "The Effect of Death Qualification on Juror's Predisposition to Convict and on the Quality of Deliberation," *Law & Human Behavior*, vol. 8, 1984, p. 53f.

of the death penalty is unconstitutional. Two United States District Courts have so found, convinced by the confluence of evidence from that series of studies, none of which by itself could have persuaded a court.[9]

SUMMARY

It is a rare event when a single piece of social science research leads to clear, unambiguous conclusions. Often, social science approaches are imperfect and require cautious interpretation. In such situations, confidence in results will be enhanced if the same results are bolstered by triangulation, that is, by a variety of independent and reliable research approaches. Depending on the issue, triangulation may make findings more precise or indeed confirm an inference about the existence or nonexistence of causal connections.

---

[9] Judge Eisele in Grigsby v. Mabry (in the Eastern District of Arkansas (1983)), and Judge McMillan in Keeton v. NC (in the Western District of North Carolina (1984)). Both decisions are on appeal. Whatever the outcome, the acceptance of the evidentiary principle of confluent social science research is noteworthy.

Author Index

Subject Index